INSECT BIOCHEMISTRY

ENCYCLOPAEDIA OF ENTOMOLOGY-I

INSECT BIOCHEMISTRY

By

Dr. M.Prakash

Dept. of Zoology
M.M.H. Post Graduate College
Ghaziabad
(U.P.)

DISCOVERY PUBLISHING HOUSE PVT. LTD.
NEW DELHI-110 002

Reprinted - 2018

First Published - 2008

ISBN: 978-93-5056-571-1 (Set)

ISBN: 978-81-8356-287-4

Insect Biochemistry

Published by:

DISCOVERY PUBLISHING HOUSE PVT. LTD.

4383/4B, Ansari Road, Darya Ganj

New Delhi-110 002 (India)

Phone: +91-11-23279245, 43596064-65

Fax: +91-11-23253475

E-mail: discoverypublishinghouse@gmail.com

sales@discoverypublishinggroup.com

web: www.discoverypublishinggroup.com

Printed at:

Infinity Imaging Systems

Delhi

Preface

Most insects are exhibitionists. They move about and engage in their daily tasks of acquiring food, courting, and reproduction undaunted by the watchful eyes of an observer. Over the span of human life on this planet people have almost certainly been fascinated and entertained by the activities of insects—what child have never stopped to admire a passing butterfly or sat and watched the coming and goings of a column of foraging ants? From such basic observations we have gained considerable insight into the biology of insects, but only recently have we begun to integrate our knowledge of behaviour, hormones, physiology and ecology into a more complete view.

The present title **Insect Biochemistry** has been carefully compiled and edited to meet the long felt needs of increasingly large number of those who have to deal with the different aspects of biology of insects in colleges, universities and research institutes. It provides a balanced and integrated treatment of the entire field of insect Biology. The title is intelligible to the educated layman but it deals with some complex ideas. It is an adequate text for all requirements in this area for most university students. Special efforts have been made to explain ideas in non-mathematical term. The primary aim throughout has been clarity, simplicity and the high standard. It will definitely prove to be a boon to teachers, students and research works in the field of Insect Biology.

The primary aim throughout has been clarity so as to make the text selective and meaningful. We hope also that general Entomologists and students of Agricultural Entomology will find in this book valuable and recent informations about the subject and also allied branch of the subject. The book provides a nucleus around which the teacher can plan a successful course without much bearing on complementary and supplementary readings. The theme of the book is novel and for the first time the various chapters in different volumes have been organised reviewing the available literature. All the topics have been illustrated with line diagrams, graphs and tables wherever necessary.

In the preparation of this book large number of books and research papers have been consulted. So no authenticity is claimed.

The author wishes to express his deepest appreciation to the many people who have contributed in one way or the other to the preparation of this title.

The author expresses his gratitude to Mr. Wasan and staff of M/s Discovery Publishing House for their whole hearted co-operation in the publication of this book.

The author tried hard to be accurate and upto date in statement and realises the impossibility of completely avoiding errors therefore, the author will greatly appreciate having his attention called to any questionable statement.

—Author

CONTENTS

INTRODUCTION

The existence of connective tissue in insects was recognized over 100 years ago but only recently with the development of new investigative techniques has its widespread distribution and significance been appreciated. Indeed, until the 1950s there was a widely held assumption that no barrier exists between the haemolymph and the tissues.

The nervous system has been prominent in connective tissue studies from the earliest beginnings. The dorsal mass of connective tissue on the abdominal connectives of adult Lepidoptera was recognized as connective tissue by Burger, D. and Nusbaum, J.. They identified it as a gelatinous structure containing cells into which the muscles of the ventral diaphragm are inserted. They also realized that it is continuous with the connective tissue around the rest of the nervous system. About this time, the sheath around the nervous system of *Oryctes nasicornis* was described by Michels, H. and later Schneider, K. named the connective tissue layer, the neural lamella, and the underlying glial cell layer, the perineurium.

Very little further work on connective tissues was reported over the next four decades. Lazarenko, T. recognized the presence of connective tissue binding the organs together in *Oryctes* larvae, but he considered that the connective tissue is produced by the breakdown of haemocytes. Similarly, Wermel, E. came to this conclusion in his work on silkmoths. Scharrer, B. published a description of the neural lamella and perineurium of *Periplaneta americana* and suggested that the perineurial cells produce the neural lamella.

Renewed interest in the connective tissues arose from the realization that the ionic concentrations in the haemolymph would block insect nerves and, therefore, the possibility that a barrier exists between the haemolymph and axons was investigated. Thus Hoyle, G. described the neural lamella of *Locusta migratoria,* but at this time the composition of connective tissue matrix was still unknown. An early birefringence study by Richards, A. led him to suggest that collagen might be present in peripheral nerves of mosquitos, and Rudall, K. reported X-ray diffraction evidence for collagen in the nerve cords of mantids, but this latter author considered that the presence of large amounts of chitin in the exoskeleton precluded the presence of significant amounts of collagen in insects.

In an attempt to characterize the neural lamella and other connective tissues, Baccetti, B. performed histochemical studies which showed that neutral glycoproteins, and possibly some acid mucopolysaccharides, are present in the connective tissue matrices of the orthopteran, *Anacridium aegyptium.* His polarized light study suggested, however, that collagen might be present, as did a similar study by Richards, A. and Schneider, D. on cockroach and silkmoth nerve cords. This was confirmed soon after with the publication of electron micrographs of periodically banded fibrils in the connective tissue matrices of several insects of different orders, and by the identification of hydroxyproline in hydrolysates of cockroach and locust nervous systems. Thus, the presence of a collagen-like protein in insects was established, but since the banding periodicities measured were very variable, there was much debate in the literature as to whether the fibrils are all similar.

From 1960 onwards there was a rapid increase in our understanding of vertebrate connective tissues, and also in the development of biochemical, histochemical and other investigative techniques. These techniques have been applied to insect connective tissues so that we can now say with confidence that insects possess connective tissue matrices of very similar composition to those of other animals, including the mammals. The discussion of the investigations which have led to this conclusion will form the major part of this chapter.

MORPHOLOGY OF THE CONNECTIVE TISSUE MATRICES

In the following discussion the connective tissue matrices found in insects will be divided into four maul groups. Very few studies, primarily concerned with the characterization of the connective tissues, have been published and thus most information on their structure is concealed in papers on other topics. In this account of the different connective tissue matrices of insects, no attempt will be made to provide a comprehensive review of the occurrence of each type; instead a guide to their identification will be given. It should be noted that the type of connective tissue associated with an organ may vary in insects of different orders.

The classification of the connective tissue matrices will be based on that currently used for vertebrate connective tissue matrices. This will differ, particularly with respect to the basement membranes, from that used by Ashhurst, D..

Fibrous Connective Tissue Matrices

Connective tissue matrices containing typically banded collagen fibrils are found in many locations. They frequently form a discrete layer lying on the outermost cells of an organ, which serves to separate the organ from the haemocoel. Layers of fibrous matrix are found around the whole nervous system (neural lamella), under the epidermis, around the fat body, reproductive system, and so on; descriptions are found in many papers. Characteristically, cells are not normally found within these connective tissue matrices, which range in thickness from less than 1 ,um to more than 10 μm.

Fibrous connective tissue matrix may be present within organs and between muscle fibres, though in rather small amounts. There are few reports of large areas of connective tissue, but a layer which is about 80 pm thick in the mature adult is found around the ejaculatory duct of the male locust, *Locusta migratoria*, and a similar, though somewhat thinner, layer occurs around the mesenteron of the adult cockroach, *Periplaneta americana.* The collagen fibrils in these layers are arranged in a random network and they contain cells, and also muscle fibres. The cells are morphologically indistinguishable from vertebrate fibroblasts.

It is usually assumed that connective tissue does not form tendon-like structures in insects. It was, however, revealed by Francois, J. that some endosternites in the head of Apterygota consist of parallel arrays of collagen fibrils. The endosternites are attached to muscles or epidermal cells, as appropriate, by hemidesmosomes. A similar, but much narrower band of connective tissue matrix joins two epidermal cells layers in the hypopharyngeal cavity of *Blaberus craniifer*. Many more structures similar to these must await identification.

The collagen fibrils in the matrices mentioned so far are clearly nded. In most descriptions a figure for the periodicity of the banding attern is given, and these vary between 55 and 70 nm. If, however, he patterns are examined carefully, it is apparent that they all conform o that of the locust and rat tail tendon fibrils are very obvious, but in sectioned fibrils bands III to X cannot be resolved clearly. It will be argued later that the periodicity of this banding pattern in all fibrils is approximately 67nm; any variations from this figure measured on electron micrographs derive from shrinkage, etc., during the preparative procedures.

The diameter of the fibrils may vary according to the age of the tissue. Thus Ashhurst, D. and Costin, N. were able to follow the development of the collagen fibrils in the sexually mature adult locust ejaculatory duct from the 0-day adult in which they are thin (about 35 nm diameter), and rather indistinctly banded, to the very large, approximately 200 nm diameter, fibrils in the mature adult, which are irregularly shaped in crosssection. It should be noted that fibrils, irregular in cross-section, can be seen in many published micrographs, and were noted in the locust neural lamella by Ashhurst, D. and Chapman, J., and in the ejaculatory duct by Martoja, R. and Bassot, J., but the latter authors considered them to be distinct from collagen. Examples of irregularly shaped fibrils occur in vertebrate tissues, e.g. in the lungs of rats.

The fibrous matrices of some orders, in particular the Coleoptera, Diptera, Lepidoptera and Protura, contain very thin, indistinctly banded fibrils. Indications of banding are occasionally seen on the fibrils and igures ranging from 15 nm to 65 nm have been suggested for the periodicity. The fibrils are collagenous, but in many tissues they are obscured by other matrix components which condense upon them, presumably during the preparative procedures. The fibrils are typically found in layers such as the neural lamella and in the large mass of connective tissue found on the dorsal side of the abdominal connectives of adult Lepidoptera. The dorsal mass contains cells derived from glial cells, which have characteristics in common with typical vertebrate fibroblasts.

It is pertinent to point out that many thin vertebrate collagen fibrils — for example those in the lens (type I collagen) and in cartilage (type II collagen) — are indistinctly banded. It appears that the number of heavy metal binding sites in a thin fibril is too small to permit sufficient stain to bind and visualize the bands clearly.

In some instances, layers of fibrous matrix may be bordered by layers of amorphous matrix. This occurs quite commonly in the Diptera; the neural lamella of *Musca domestica* has a 60 nm thick amorphous layer peripheral to the fibrillar region and several regions of connective tissue in the rectum of *Calliphora erythrocephala* have alternating amorphous and fibrillar layers. A very thin amorphous layer may be found on the surface of the neural lamella of locusts and cockroaches. The precise nature of this amorphous material is unknown. It may simply be a layer of non-collagenous proteins and glycosaminoglycans, or where it is a relatively greater component of the tissue, it could also contain a non-fibrous collagen.

The connective tissue around the ejaculatory duct of *Locusta*, was classified as cartilage by Martoja, R. and Bassot, J. because in their preparations the cells appeared to be in lacunae. Subsequent work on this layer failed to substantiate this classification. There are no other suggested locations of cartilage in insects.

Basement Membranes

Most vertebrate cells, with the exception of fibroblasts and blood cells, produce a basement membrane. This can be seen in light microscope preparations as a dense line under the basal membrane of an epithelial cell, or around a muscle cell; it is especially clear after the periodic acid-Schiff reaction since the basement membrane contains glycoproteins. In electron micrograph sections, the basement membrane is seen to consist of two layers: the electron-lucent lamina rara (or lucida) which is about 50 nm thick, adjacent to the cell membrane, and the lamina densa, which appears amorphous, though some very fine filaments can often be discerned within it. The thickness of the lamina densa is variable, but it is not usually more than 100 nm thick. It contains type IV collagen. Below the lamina densa there may be a dense meshwork of collagen fibrils, which has been called the reticular layer. This is produced by fibroblasts and not by the epithelial cells, and is not strictly part of the basement membrane. The term "basement membrane" is usually used to include the laminae rara and densa, but some workers restrict the term to the lamina densa.

Basement membranes with this typical structure are found under many insect epithelia, around muscle fibres, and around fat body cells; there is considerable variation in the thickness of the lamina densa. The basement membrane separates the muscle fibres from the haemolymph, but in other organs the basement membrane is adjacent to fibrous connective tissue. This situation is seen in the mesenteron of the cockroach.

A typical basement membrane may appear as a stage in the development of a connective tissue matrix. In *Schistocerca* embryos, for example, the nervous system is initially surrounded by a basement membrane, but as more matrix is laid down the dense layer thickens and the lamina rara is obliterated.

Thick Non-Fibrous Layers of Connective Tissue Matrix

There are many instances of layers of connective tissue matrix, which may be several micrometres thick, but which do not contain fibrils: they are also closely apposed to the adjacent cells. Thus, by virtue of their thickness and the absence of a lamina rara, they cannot be considered as basement membranes.

The occurrence and location of this type of matrix varies from insect to insect. In *Rhodnius prolixus* the epidermal cells lie on a matrix layer of this type. Several endocrine organs of the cockroach *Leucophaea maderae* are surrounded by sheaths of this type of matrix which also penetrates between the cells, and similar layers lie under the Malpighian tubules of the housefly, *Musca domestica*. These are only a few examples to illustrate the wide distribution of thick,

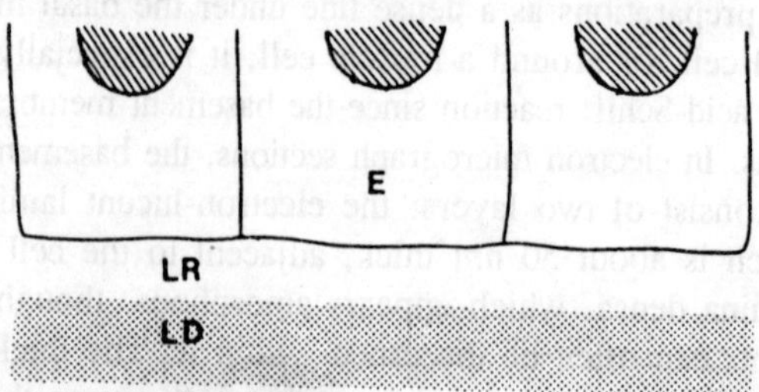

Fig. 1.1. A diagrammatic representation of a basement membrane. The lamina rara (LR) separates the lamina densa (LD) from the epithelial cells (E).

apparently non-fibrous layers of connective tissue matrix and it is noteworthy that typical fibrous matrices occur in all these insects, particularly in the neural lamella. Small areas between layers of cells are sometimes filled by this matrix. Recently, the so-called "basement membrane" of the compound eye of several insects has been examined. It consists of a layer of amorphous or finely fibrillar connective tissue matrix which, in different species, may vary from 1 um to 40 μm in thickness and contains cell processes; the term basement membrane is clearly a misnomer in this instance.

Matrix, very similar to that described here, has been found forming connecting strands or ligaments between the heart, pericardial cells and other structures in Diptera and Lepidoptera; elastic fibres occur in these strands. The composition of this connective tissue matrix is unknown. One presumes that'a collagenous protein is present, but as yet this cannot be determined. It is very difficult to detect the presence of glycosaminoglycans histochemically, as most of the layers are less than 1 μm thick. Thus many questions remain to be answered about this type of matrix.

Specialized Connective Tissue Layers

Bertram, D. and Bird, R. described a complex layered structure lying under the midgut epithelial cells of *Aedes aegypti*. Examination of longitudinal and tangential sections revealed that it consists of three or four layers, each of which is an array of cylindrical units with less electron-dense centres joined to form a network, which is expanded and distorted after a blood-meal. A similar structure has been seen in the midgut of fleas, but here the units are not joined together.

Grid-like layers were found under the midgut epithelium of six of the 13 species of Coleoptera examined by Holier, P.. He found a single layer which appears in longitudinal sections as a ribbon-like structure transected by dark and light bands. In tangential sections, two types of units are distinguished; small electron-dense units are arranged around larger less dense units, and are held together by bridges. In *Oryctes* larvae the midgut epithelium lies on a three-layered structure. Here the inner layer resembles that described by Holter, but the lighter units increase in size in the middle and again in the outer layer.

A further type of modified layer is found in the midgut of the Heteroptera. In *Nepa cinerea* and *Ranatra linearis*, a typical lamina densa is separated from the cell membrane by an electron-lucent layer 35 nm thick and a second discontinuous layer, 90 nm thick, which consists of a series of polygonal plaques, separated by gaps of about 30 nm and bridged by thin filaments.

The chemical nature of these layers is unknown. The dense plaques and outer layer in *Ranatra* are digested by collagenase, but the layers in *Oryctes* are resistant. The layers in *Ranatra, Harpalis rufipes* and *Oryctes* are positive with the periodic acid-Schiff (PAS) test, but only in *Oryctes* are the layers positive with tests for glycosaminoglycans. Thus, while the positive PAS reaction suggests that these layers contain glycoprotein, their other constituents are yet to be determined.

The function of these networks is not understood; indeed Gouranton, J. considers it is impossible to predict. The only clue might come from the mosquitoes and fleas in which the midgut is very distended after a blood-meal; the network might permit the stretching necessary to accommodate the food.

Similar structures were described around the ovarioles of *Aedes aegypti*, but no more examples have been recorded so far in other insects; recently, modified layers have been found under the midgut epithelia of two Crustacea.

CONSTITUENTS OF CONNECTIVE TISSUE MATRICES

Connective tissue matrices are very diverse, since they range between the hard, dense, calcified matrix of bone and the soft, jelly-like vitreous of the eye.

Nevertheless, they are composed of the same molecules: the fibrous proteins, collagen and elastin, glycosaminoglycans and many glycoproteins.

The collagens from animals of many phyla have been studied extensively. At the present time, ten types of collagen molecule extracted from vertebrate tissues have been characterized, and yet more are under investigation. Those invertebrate fibrous collagens investigated appear to be very similar to type I collagen, or type I trimer. In addition, basement membrane collagens are present in invertebrates, but their similarity to mammalian type IV collagen is not fully established.

The collagen molecule, which is approximately 300 nm long and 1.5 nm in diameter, consists of three polypeptide chains, or Lx-chains. Each molecule has a central triple helical region, or domain. Each a-chain is coiled into a left-handed helix with about three amino acids per turn and the three a-chains of the molecule are then coiled around each other to form the right-handed superhelix; the short terminal regions are not triplehelical. The a-chains may be identical, or one may be dissimilar.

Each α-chain consists of just over 1000 amino acid residues. In the helical domain every third amino acid is glycine, and hence glycine accounts for approximately one-third of the total amino acids of a collagen. Glycine is the smallest amino acid, and in the collagen triple helix glycine residues occupy the restricted space in the centre of the helix. The collagen molecule is also unusual since about 20% of the amino acids are proline, or hydroxyproline. These amino acids are important for the triple helical conformation: the imino acids (proline and hydroxyproline) confer a polyproline-type helix on the α-chain, while glycine stabilizes the triple helix by forming inter-chain hydrogen bonds. The hydroxylation of proline to form hydroxyproline occurs after the assembly of the a-chain, and its presence is essential for the stability of the triple helix at body temperatures. Hydroxyproline has been found in very few other proteins, notably vertebrate elastin, the Clq component

TABLE 1.1: SOME DIFFERENT TYPES OF VERTEBRATE COLLAGENS

	α-chain composition	*Fibrils*	*Typical locations in vertebrates*
Type I	$\alpha 1(I)_2\alpha 2$	thick	skin, bone, tendon, arterial wall, uterus.
Type I trimer	$\alpha 1(I)_3$	thick	skin
Type II	$\alpha 1(11)_3$	thin	cartilage
Type III	$\alpha 1(III)_3$	thin	arterial wall, uterus, skin
Type IV	$\alpha 1(IV)_2\alpha 2(IV)_{3i}$ $\chi 1(IV)_3$ or $\alpha 2(IV)_3$	non-fibrous	basement membranes
Type V	$\alpha 1(V)_2\alpha 2(V)$	non-fibrous?	placenta, blood vessels

of complement and the tail region of acetylcholinesterase. The remaining amino acids in the polypeptide chain are variable, but noteworthy among them is lysine. Some lysine residues are hydroxylated post-translationally to form hydroxylysine. Both lysine and hydroxylysine may be glycosy-lated and some residues are involved in the formation of the cross-links, dehydrodihydroxylysinonorleucine and dehydrohydroxylys-inonorleucine, upon which the stability of the molecules and fibrils depends. The number of lysine and hydroxylysine residues is thus significant.

The molecules of the fibrous forms of collagen come together in a very precisely determined manner to form the characteristically banded fibrils. The repeating period (D) is approximately 67 nm; this figure has been determined by X-ray diffraction of native, hydrated, rat tail tendon collagen. Each molecule is 4.4 D in length. In the fibril the molecules are aligned in parallel but each is staggered by exactly 67 nm (D) with respect to its neighbours. The gaps between the ends of the molecules are an intrinsic feature of this model, which is the modified quarter-stagger model of Hodge, A, and Petruska, J.. The banding pattern seen in electron micrographs after both positive and negative staining is a reflection of the distribution of the polar, or charged, amino acids along the α-chains.

The vertebrate type IV basement membrane collagens do not form fibrils *in vivo*. The molecules are more diverse and are longer than those of the fibrous collagens. It is suggested that the longer length may be due to the retention of the C-terminal extension peptide and that this may be responsible for the knob seen in electron micrographs at one end of the molecule. The molecule consists of at least two dissimilar polypeptide chains, with molecular weights between 165,000 and 185,000. A model for the aggregation of type IV collagen molecules *in vivo* has been proposed recently. It is suggested that two molecules are joined at the C-terminal ends to form a dimer and that each N-terminal end of this dimer is linked to the Nterminal ends of three more dimers, so that a network of collagen molecules is produced.

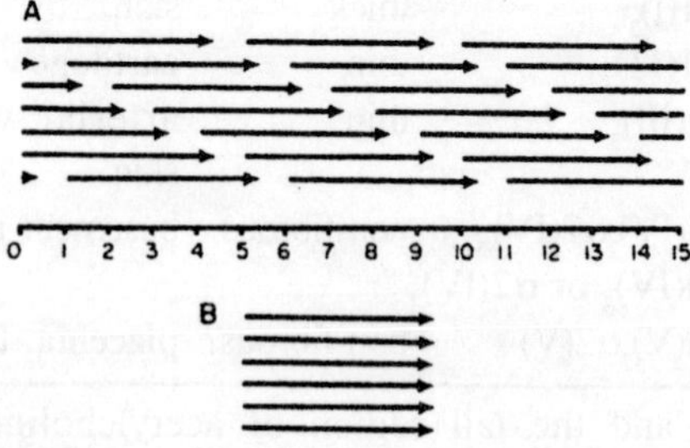

Fig. 1.2. A: Diagram to show the precise arrangement of the collagen molecules in a fibril according to the modified quarter-stagger model of Hodge.

The only invertebrate basement membrane collagen to be studied extensively to date is from the intestinal basement membrane of the roundworm, *Ascaris suum*, and it was concluded that this invertebrate collagen is similar to vertebrate basement membrane collagens.

The glycosaminoglycans are a more heterogeneous group of substances. They are long-chain polymers composed of repeating, identical disaccharide units. Each disaccharide consists of an amino sugar, glucosamine or galactosamine, and a uronic acid moiety, either glucuronic or iduronic acid, or in some instances a sugar. The polymers have anionic charges along their length; these are the carboxyls of the uronic acid, and sulphates which may be attached to the amino sugar. The glycosaminoglycans include hyaluronate, chondroitin-4- and chondroitin-6sulphates, dermatan sulphate, keratan sulphate, heparin and heparan sulphate; the repeating disaccharides. With the exception of hyaluronate, all the glycosaminoglycans have an obligatory association with a polypeptide, to form a proteoglycan. The glycosaminoglycan is attached to the polypeptide via a linkage region which consists of a trisaccharide unit (Gal-GalXyl); the xylose residue is glycosidically linked to a serine residue. Typically, the polymers of chondroitin sulphate are much longer than those of keratan sulphate. The size and composition of the proteoglycans varies with the tissue.

In cartilage the proteoglycans themselves are attached by a link protein to a single strand of hyaluronate, and the resulting aggregate, which in electron micrographs appears to have a feather-like form, may be several micrometres in both length and breadth. Heparan sulphate is associated with cell surfaces. Apart from one or two rare examples, heparin is the only intracellular glycosaminoglycan and it occurs in the mast cell granules of vertebrates; it is released during trauma and is an anticoagulant of the blood. Hyaluronate may form long polymers more than 1 μm in length and its physical conformation depends on the ionic strength of the surrounding medium.

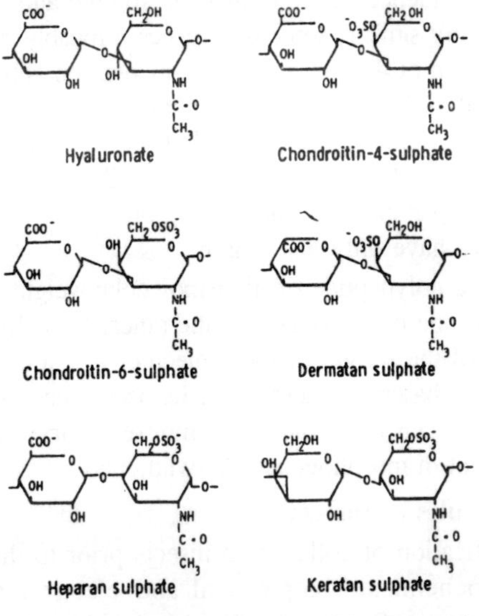

Hyaluronate

Chondroitin-4-sulphate

Chondroitin-6-sulphate

Dermatan sulphate

Heparan sulphate

Keratan sulphate

Fig. 1.3. The repeating disaccharide units of the glycosaminoglycans.

The physical properties of the glycosaminoglycans are related to their size and anionic charges. Because they form large polymers and aggregates and have a high charge density, they are highly expanded molecules which occupy a significant volume of the matrix. Thus it has been shown by Laurent and his co-workers that they form charged sieves which can impede large molecules, and repel other anionically charged ions and molecules which try to cross the matrix. The chondroitin sulphates are normally found together with fibrous collagens and may also affect the polymerization of collagen molecules to form fibrils. Another important property of glycosaminoglycans is their ability to bind water. The viscosity of hyaluronate solutions is determined by the amount of water bound, and its ability to form viscous solutions is important for its lubricating and shock-absorbing function in, for example, joints and the vitreous humour of the eye.

The glycoproteins of connective tissue are very heterogeneous, and have received far less attention than the collagens and glycosaminoglycans. Recently, two glycoproteins from vertebrate tissues, fibronectin and laminin, have been characterized. Fibronectin is found in a soluble form in the blood and plasma, but in an insoluble form on cell surfaces, and while both forms have been studied extensively, it was realized only recently that only one molecule exists. The fibronectin molecule consists of two identical polypeptide chains, and while immunological evidence suggests that the soluble and insoluble forms are chemically very similar, some differences, probably due to variations in the glycosylation of the molecules, are evident on gel electrophoresis. It is suggested that fibronectin might form a bridge between the cells and matrix, since the molecule has binding sites for collagen, glycosaminoglycans, fibrinogen, actin and cell surfaces. Thus, it may be important in developmental and healing processes, but investigations along these lines have only just begun.

Laminin is a polypeptide of high molecular weight which consists of two subunits, one twice the size of the other. Carbohydrate accounts for about 15% of the weight of the molecule. It is a constituent of the lamina rara of the basement membrane. Its occurrence, as demonstrated by immunofluorescence, in the early mouse embryo suggests that it may be important in morphogenetic movements.

Collagen Molecules of Insects

The identification of collagen in insects prior to the development of modern biochemical, biophysical and electron microscopical techniques was very difficult. The first biochemical studies used crude paper chromatography from which the presence of hydroxyproline, and hence collagen, was established. The next attempt to confirm the biochemical similarity of insect and other collagens was by Harper, *E. et al.*; their amino acid analyses of collagen from several cockroaches confirmed that a true collagenous protein is present.

In the last 15 years many new methods and parameters for the characterization of collagen molecules have been developed. Thus, the question which is asked about any collagen molecule is: is it similar to one of the well-established vertebrate collagen types, or is it a

TABLE 1.2: AMINO ACID COMPOSITION OF MAMMALIAN AND INSECT COLLAGEN A-CHAINS.

Amino acids	Amounts of amino acids (residues per 1000)					
	Human skin al (I) chains 2.5 M NaCl precipitate	*Human skin al (I) chains 4.0 M NaCl precipitate*	*L. migratoria 4.0 M NaCl precipitate*	*P. americana 4.0 M NaCl precipitate*	*P. americana*	*L. migratoria basement membrane (calculated)*
Hydroxyproline	93.0	98.0	109.3	107	41.2	90.3
Aspartic acid	46.3	45.4	46.7	50	66.4	83.0
Threonine	20.6	18.0	24.1	24	40.1	23.7
Serine	36.2	37.3	32.9	39	62.5	23.6
Glutamic acid	62.8	66.6	92.2	91	86.5	76.8
Proline	131.0	126.1	129.8	120	88.1	67.8
Glycine	320.1	326.2	324.8	320	235.2	300.7
Alanine	115.6	127.1	49.0	78	104.7	33.4
Valine	21.6	18.3	16.3	22	44.0	26.2
Methionine	7.3	7.6	n.d.	n.d.	3.8	n.d.
Isoleucine	9.6	9.5	18.5	16	25.3	42.8
Leucine	26.4	24.1	38.2	35	35.3	56.2
Tyrosine	2.3	1.9	1.6	n.d.	31.7	11.6
Phenylalanine	12.8	11.9	10.4	13	27.6	20.5
Hydroxylysine	4.6	9.6	24.1	16	26.4	64.6
Lysine	31.9	25.2	14.1	20	20.7	16.9
Histidine	3.9	3.6	4.6	4	18.2	0.17
Arginine	54.7	50.0	53.1	45	34.5	55.1

unique molecule? This question must be approached in a variety of ways. Firstly, it is possible to discover by electron microscopical examination of native fibrils and segment-long-spacing (SLS) crystallites whether the molecular lengths and distribution of polar amino acids along the a-chains are similar. Secondly, biochemical evidence from cyanogen bromide peptide patterns, amino acid analyses, salt precipitation and other techniques gives information which can be compared to that of other collagens.

The first problem encountered in attempting such an investigation of insect collagen is getting sufficient collagen from a restricted source. Ashhurst, D. and Costin, N. had used the ejaculatory duct of adult male locusts (*Locusta migratoria*) for several investigations, and since the amount of connective tissue around the duct is relatively large, this source of collagen was used for extensive studies.

Invertebrate collagens are notoriously insoluble; in order to reduce the cross-linking of the collagen, newly moulted adult locusts were fed on wheat seedlings sprayed with β-aminoproprionitrile, a lathrytic agent which prevents the formation of cross-links. After 6 days the ducts were removed, and acid-soluble collagen was extracted. By dialysing an acid extract against either ATP at acid pH, or phosphate buffer at pH 9.2, it is possible to form either SLS crystallites or reconstituted native fibrils, respectively. These may then be compared with similarly prepared crystallites and fibrils made from vertebrate collagens. Micrographs of positively stained SLS crystallites and reconstitued fibrils made from locust collagen are seen in Figs 21 and 22 together with those made from rat tail tendon, type I collagen. It is immediately apparent that the banding patterns of both SLS crystallites, and both reconstituted fibrils, are identical. A similar investigation by Francois, J. *et al.*, using pepsin-solubilized collagen from the mesenteric sheath of the cockroach *(Periplaneta americana)* yielded exactly the same result. Thus, the following conclusions can, be drawn from the SLS crystallites: firstly, the locust and cockroach collagen molecules are the same length as the rat tail tendon molecules, and secondly, the distribution of polar amino acids along the a-chains is the same as that in mammalian type I molecules. The molecules form fibrils with a periodic banding pattern exactly similar to that of mammalian type I collagen, which means that when the molecules are assembled into fibrils, the molecular arrangement must be the same. Hence it follows that the periodicity of the insect fibrils must be identical to that of mammalian fibrils, that is 67 nm. Any variation from this figure measured on electron micrographs must be due to the preparative techniques. The details of the banding pattern are not seen so readily in sectioned fibrils, but bands I and XI can usually be resolved, while bands III to X are less clear.

The amino acid analyses of several insect collagens and mammalian type I and type I trimer. While amino acid analyses are somewhat variable, it is immediately apparent that the composition of insect collagens is similar to that of the mammalian collagens. The significant features are that in all, glycine accounts for approximately one-third of the total amino acids, hydroxyproline is present, and together with proline, makes up about 20% of the amino acids. Hydroxylysine occurs in amounts greater than in mammalian type I, but in type I trimer from human skin, the number of hydroxylysine residues is also raised.

The locust and cockroach collagen molecules consist of one type of α-chain - that is, they are trimers. When the a-chains in the pepsin digest were isolated by salt fractionation, the locust material produced a minor precipitate at 2.4 M NaCl, and a major precipitate at 4.0 M NaCl. All the cockroach Lx-chains precipitated at 4.0 M NaCl. The precipitates were analysed by sodium dodecylsulphate/acrylamide gel electrophoresis; both the locust and cockroach 4.0 M NaCl precipitates produced a single band with the same mobility as the al(I) chain. This shows that only one type of a-chain is present and that it is similar to that of the mammalian type I α1 chain. The 2.4 M NaC1 locust precipitate after mercaptoethanol treatment gave a faint band with a mobility greater than the major component. It is thought that this minor component is a basement membrane collagen, and it will be discussed later.

Another criterion on which collagens may be compared is the pattern of peptides produced after cleavage of the polypeptide chains at the methionine residues by cyanogen bromide. Neither the locust, nor the cockroach collagen has a pattern comparable to that of the al (I) chain, but this indicates only that the position of the methionine residues differs. Sea anemone collagen, which has many properties in common with mammalian type I collagen, also has a different spectrum of CNBr peptides.

The molecular cross-link pattern of the locust collagen after borohydride reduction reveals that this collagen possesses the two reducible cross-links, dehydrodihydroxylysinonorleucine and dehydrohydroxylysinonorleucine; the cockroach collagen apparently possesses only the former. These crosslinks are found in mammalian and other invertebrate collagens. They are formed by lysine and hydroxylysine residues, and thus the increased number of hydroxylysine residues in these and other invertebrate collagens is probably a significant factor in their greater insolubility wh .n compared to vertebrate collagens since the dihydroxy cross-link is more stable than the lysine derived monohydroxy cross-link.

Francois, J. and his co-workers also found that approximately 8.6% of the total dry weight of cockroach collagen is neutral sugar, which is consistent with the general finding that invertebrate collagens contain more hexoses Than the vertebrate molecules.

On the basis of their data Ashhurst, D. and Bailey, A. concluded that the fibrous collagen molecule extracted from *Locusta* is very similar to that of vertebrate type I trimer collagen, differing only in the distribution of some amino acids. Despite many obvious points of similarity between the cockroach and type I trimer collagens, Francois, J. *et al.* considered, on the basis of the amino acid composition, that this collagen may be more similar to mammalian type II, i.e. the cartilage collagen. Their argument is based solely on the higher glutamic acid and hydroxylysine content of the cockroach collagen, and ignores their other significant evidence, which is firstly that the SLS crystallite banding pattern is the same as that of type I collagen, while type II has a completely different pattern, and secondly, that the cockroach collagen elutes on CM-cellulose chromatography later than the al (I) chains, while type II a-chains elute with the type I a-chains. Thus, they base their conclusion on the relative amounts of two amino acids, and choose to ignore the weight of their other evidence. It would indeed seem peculiar if the cockroach collagen is so unique among invertebrate collagens.

The thin, indistinctly banded fibrils of the Lepidoptera and other orders have been an enigma. Some work currently in progress has shown that pepsin-solubilized collagen from the dorsal mass of . the nerve cord of late pupae of *Manduca sexta* forms SLS crystallites similar to those of type I collagen and that a1-chains, with the same mobility as al (I)chains on sodium dodecylsulphate/acrylamide gels are precipitated by 2.4 M NaCl (unpublished observations). These results indicate that the moth collagen may be very similar to other insect collagens.

Several other invertebrate collagens, from animals such as sea anemones, tapeworms, liver flukes, earthworms (body wall collagen), crabs, various molluscs, sea cucumbers and ascidians have been identified as trimers with al-chains similar to those of type I collagen. A few instances of both al and 12 chains in the collagens of crustaceans and molluscs have recently been documented. While the amount of data on each collagen varies, all that is available is in agreement with the conclusion that the fibrous collagens of the invertebrates have many properties in common with vertebrate type I collagen, and in some instances, more especially with type I trimer. Thus, it appears that a molecule similar to the type I trimer already existed in the coelenterates and that during the evolution of the higher invertebrates and vertebrates this molecule has been conserved with only minor alterations in some of the amino acids.

The basement membrane collagen of the locust is more elusive as it is present in such small amounts. It was detected on the gels after mercaptoethanol treatment, which breaks disulphide bonds, as a faint band with a mobility greater than the major component of the 2.4 M NaCl fractional precipitate. Its amino acid composition was calculated. The mobility of this component on dodecylsulphate/ acrylamide gel electrophoresis is similar to that of al (IV) chains and certain features of the amino acid analysis, for example, the high lysine/ hydroxylysine ratio and the high contents of hydroxyproline and leucine, suggest that it is a basement membrane collagen. More recently a collagen has been isolated from locust leg muscle, which has similar properties to the basement membrane collagen from the ejaculatory duct; work is in progress to characterize this collagen further (Ashhurst, Bailey and Sims, unpublished observations).

A collagen from *Drosophila,* with morphological and biochemical properties similar to those of mammalian type IV collagen, has recently been characterized by Fessler, J. and his co-workers. The collagen has been extracted from both the media of cultured *Drosophila* KC cells and from *Drosophila* larvae. Electron micrographs of rotary shadowed isolated molecules show a molecule, approximately 465 nm long with a knob at one end. This appearance and size is similar to that of mouse type IV procollagen, and there is evidence that the knob is at the C-terminal end of the molecule. Antibodies to this molecule are bound by the basement membranes of *Drosophila* larvae.

Another approach, used recently by Monson and her co-workers was to screen a library of *Drosophila melanogaster* genomic DNA fragments with a chicken proa2(I) cDNA clone to isolate collagen-like genomic clones. The fragment isolated is composed of two large coding sequences which together specify a sequence of 469 amino acids. The polypeptide is composed almost entirely of the Gly-X-Y repeat characteristic of the helical portion of the molecule, but there are four minor interruptions in the sequence. Similar interruptions of the helical portion of the molecule occur in mouse basement membrane collagen, and hence it is suggested that this *Drosophila* collagen gene may encode part of a non-fibrous, basement membrane collagen molecule.

Thus evidence for the presence of insect basement membrane collagens, with properties similar to those of the mammalian type IV collagens, is gradually accumulating.

Glycosaminoglycans

Most information about the glycosaminoglycans of the connective tissue comes from histochemical studies. Baccetti, B. found that the neural lamella and the connective tissue matrices associated with the fat body, gut, Malpighian tubules, muscles and epidermis of the orthopteran *Anacridium aegyptium* are positive with the periodic acid-Schiff (PAS) test, but that only the matrices of the nervous system, fat body and gut stain metachromatically with toluidine blue. The connective tissue matrices of lice and the neural lamellae of the locust, *Locusta migratoria,* and cockroach, *Periplaneta americana,* appeared to differ in that they did not stain metachromatically with toluidine blue. This led to the conclusion that acidic carbohydrates were not universally present in insect connective tissue matrices. This finding seemed anomalous, since glycosaminoglycans are usually associated with fibrous connective tissues; collagen fibrils were identified in neural lamellae in 1958. A little later it was found that the connective tissue mass on the dorsal side of the abdominal nerve cord of the waxmoth, *Galleria mellonella,* is strongly metachromatic with toluidine blue, and that this metachromasia is hyaluronidase-labile.

More critical histochemical tests for glycosaminoglycans were developed during the 1960s by Scott and Spicer and their co-workers. These techniques enable the identification of the individual glycosaminoglycans. Thus when this battery of tests was performed on the neural lamellae of several adult insects of different orders *(Carausius morosus, Galleria mellonella, Locusta migratoria, Periplaneta americana)* the results in Table 4 were obtained. They suggest that all are similar in possessing chondroitin, and possibly dermatan, sulphates, but while those of the stick insect, locust and cockroach contain a highly sulphated hyaluronidase-resistant component, identified as keratan sulphate, the neural lamella and dorsal connective tissue mass of the waxmoth nervous system does not. The amount of keratan sulphate present usually increases as a matrix ages and as the connective tissue of the adult waxmoth develops over a few days and the adult lives only a short time, this ageing process may not occur. A recent study of the neural lamella during larval development in *Locusta migratoria* and *Periplaneta americana* showed that sulphated glycosaminoglycans increase during development and that keratan sulphate is present only in the last larval instar and adult. The presence of glycosaminoglycans in the neural lamella of the waxmoth was recently confirmed by ultrastructural studies using ruthenium red.

The connective tissue matrices associated with chemical tests. Thus, the connective tissues associated with the ejaculatory duct of the male locust (*Locusta migratoria*) and with the mesenteron of the cockroach (*Periplaneta americana*) contain increasing amounts of sulphated glycosaminoglycans as they develop through the larval instars to the adult, when chondroitin and keratan sulphates are present.

Thus the anomaly of fibrous matrices seemingly devoid of chondroitin and keratan sulphates was resolved. The dye alcian blue is most commonly used to localize glycosaminoglycans; it is pertinent to reiterate here that alcian blue at pH 2.5, i.e. in 3% acetic acid, is often not bound by glycosaminoglycans due to interference by the polar groups on the proteins although it is in these conditions that the dye is most commonly used. Thus a battery of tests is essential before any valid conclusions can be drawn.

Histochemical tests locate the glycosaminoglycans in the connective tissues, but unfortunately they do not provide conclusive proof of the presence of a particular substance. This can only be done biochemically, but so far few attempts have been made. In early experiments, hyaluronate, which was presumed to be from the glial lacunar system, was found in extracts of cockroach ganglia and Estes, Z. and Faust, R. concluded that only hyaluronate is present in the midgut of *Galleria mellonella* larvae. Sharief, F. *et al.* isolated hyaluronate, chondroitin sulphate and heparan sulphate from whole embryos and larvae of the fly, *Phormia regina,* and confirmed the presence of sulphate groups autoradiographically. They attempted to locate the glycosaminoglycans histochemically using alcian blue, pH 2.5 and colloidal iron, and found that the connective tissues associated with the gut, nervous system, fat body and salivary glands gave positive reactions. The positive reactions of other parts of these organs may be attributable to mucins, with either sialic acid carboxyls or sulphate groups, or in the case of the nuclei, to the phosphate groups on DNA. Alcian blue and other cationic dyes are bound by any anions in the tissues, not just those on the glycosaminoglycans.

A more thorough biochemical analysis of the glycosaminoglycans in the fly, *Calliphora erythrocephala,* during development from the larva to the adult fly was performed by Hoglund, L..

The extracted glycosaminoglycans were fractionated and the amounts of hexosamine, hexuronic acids, neutral sugars, sialic acid, etc., in the eluted fractions were determined. The fractions were also subjected to electrophoresis and infrared spectroscopy and enzyme lability was investigated. Thus, hyaluronate, chondroitin, and chondroitin sulphate, and possibly heparan and keratan sulphates were identified in the adult fly. Hyaluronate, chondroitin, and heparan sulphates, together with a "larval acid mucopolysaccharide" ("larval AMPS") were found in the embryo from late cleavage onwards and in the larva. The "larval AMPS" is not sulphated, nor is it labile to testicular hyaluronidase. It contains galactosamine, glucosamine, hexuronic acid and galactose, and Hoglund suggests that it may be a mixture of components.

No attempt was made to determine the tissue localization of the glycosaminoglycans in *Calliphora*. These biochemical studies do, however, confirm that glycosaminoglycans are present in insect tissues from an early stage in development, and that the sulphated glycosaminoglycans, chondroitin and keratan sulphates, more particularly the latter, occur in increasing quantities as the adult stage approaches.

The discussion of the glycosaminoglycans so far has been confined to those in the fibrous matrices. It is only recently that glycosaminoglycans have been detected in mammalian basement membranes, and similarly observations on insects are recent. Francois, J. reported that the basement membrane of the epithelial cells of the mesenteron of *Periplaneta* binds alcian blue under conditions which indicate the presence of sulphated glycosaminoglycans, and Dutkowski, A. found that the basement membrane of the fat body of *Galleria* binds ruthenium red, which again suggests the presence of glycosaminoglycans.

In her early study on cockroach ganglia, Ashhurst, D. found that the contents of the glial lacunar system stain metachromatically with toluidine blue, and this material was identified biochemically as hyaluronate.

There are suggestions in the literature that glycosaminoglycans may be found in other locations. Marshall, A. characterized the mucus in the dwelling tubes of cercopoids and found evidence for glucosamine and glucuronic acid. Similarly, extracts of the peritrophic membrane of *Bombyx mori* contain glucosamine and glucuronic acid. There are also many examples in which it is suggested on the basis of limited alcian blue staining that glycosaminoglycans are present in cells. It should, however, be borne in mind that glycosaminoglycans are rarely found within cells and that mucoproteins, i.e. mucus, may be acidic due to sialic acid or sulphate groups attached to the molecule.

Glycoproteins

The glycoproteins of vertebrate connective tissues have received little attention until recently and nothing is known of the biochemistry of the glycoproteins in insect connective tissue matrices. That glycoproteins are present is shown by the strongly positive reaction with the standard PAS test. The glycosaminoglycans do not give a positive reaction with the standard PAS test, and so another group of carbohydratecontaining substances is responsible for the positive reaction.

That insect connective tissue matrices of different types are universally PAS-positive was observed in all the early histochemical studies and this has been confirmed by all the subsequent studies. Since collagen is glycosylated this may account for some of the PAS-positivity, but additional glycoproteins, as yet unidentifed, must be present. It is interesting, however, that fibronectin (LETS protein) could not be found among the cell surface proteins isolated from four lines of dipteran and lepidopteran cells in culture. Laminin has, however, been isolated from *Drosophila;* the molecule has a similar appearance in electron micrographs to that of vertebrate laminin, and its sedimentation properties and amino acid analysis are also similar. Further work on insect connective tissue glycoproteins must await the development of suitable experimental techniques.

FORMATION OF CONNECTIVE TISSUE MATRICES

The constituents of connective tissue matrices are synthesized and secreted by the cells on which they lie, or by fibroblasts within the matrix. Vertebrate fibroblasts produce collagen, glycosaminoglycans, elastin and the microfibrillar protein, and glycoproteins. A typical fibroblast is a spindleshaped cell which has a large amount of rough endoplasmic reticulum with very dilated cisternae containing an electron-dense amorphous material. The cisternae may come very close to the cell membrane and lose the ribosomes on the cisternal membrane adjacent to the cell membrane. The Golgi complexes are rather small and usually consist of many small vesicles with no specific arrangement and only a few lamellae. There are no secretory granules in fibroblasts; the newly synthesized matrix molecules are not stored, but secreted straight into the matrix. A few lysosomes may be present. Insect fibroblasts share these features in common with vertebrate fibroblasts.

Some cells have other functions and produce connective tissue substances only at certain periods. Thus, all cells produce their own basement membrane and smooth muscle cells produce the connective tissue which surrounds them. The situation is exactly the same in insects; the perineurial cells produce the neural lamella, the epidermal cells produce the connective tissue matrix on which they lie, the underlying cells produce the tunica propria of the ovariole wall, and so on. Evidence for this derives from the presence of large amounts of dilated rough endoplasmic reticulum in the cytoplasm during periods of active matrix secretion.

The contribution of haemocytes, if any, to the connective tissues is most probably, as suggested by Whitten, J., limited to the provision of nutrients to the cells responsible for their synthesis, though it has been suggested by several workers that haemocytes contribute directly to the insect connective tissue matrices. The evidence is based on the proximity of haemocytes to developing connective tissues and the PAS-positivity of the connective tissue matrices and granules within the haemocytes, or the presence of dilated rough endoplasmic reticulum in the blood cells. The evidence is at best circumstantial, and does not take into account our knowledge of the biochemistry and mode of formation of connective tissues in other animals. It has been evaluated fully elsewhere and will not be considered further here.

Embryonic Development

Because they surround organs and bind the cells together, connective tissue matrices are usually formed late in development, although as will be discussed later, their importance in early stages of embryogenesis is becoming increasingly apparent.

Only one study of connective tissue formation in an embryo has been attempted, and this was confined to the neural lamella of the locust *Schistocerca gregaria*. In conditions in which embryonic development lasts for 12 days, the neural lamella is first seen as a basement membrane in the 9-day embryo. During the next 3 days more matrix is laid down, so that a typical, thin, fibrous neural lamella is present in the newly hatched nymph. During this period the perineurial cells enlarge and contain typical rough endoplasmic reticulum, but this is much reduced when the firstinstar neural lamella is fully developed.

The neural lamella gradually thickens during the larval instars, but no-one has yet examined the perineurial cells to see if there is renewed development of rough endoplasmic reticulum at, forexample, the time of moulting when active matrix secretion might occur. The perineurial cells of firstinstar larvae *of Galleria mellonella* contain a lot of rough endoplasmic reticulum, while it is reduced in the second and third instars; unfortunately the larvae were not accurately aged within each instar and so it is impossible to conjecture whether the second and third instar cells might be in a quiescent intermoult period. There is evidence, however, that new endoplasmic reticulum might be formed in the perineurial cells of the tick, *Boophilus microplus,* after the adult moult.

Post-Embryonic Development

It is obvious from the preceding section that the connective tissues laid down in the embryo continue to develop in exopterygotes until the adult stage is reached, or in endopterygotes, until pupation.

The connective tissue around the mesenteron of the cockroach, *Periplaneta americana,* is present in the first instar as a layer 6-101.lm thick, in which active fibroblasts, muscle cells and nerves are embedded in a fibrous matrix. A basement membrane separates the epithelial cells from the connective tissue. In the succeeding instars, the layer gets thicker to reach the adult dimensions of 10-30 /An. The cellular content of the layer is similar throughout, and both active and inactive fibroblasts were observed in the adult. The initial stages of the development of this layer are in the embryo.

The connective tissue around the ejaculatory duct of the male locust, *Locusta migratoria,* in contrast, is formed during post-embryonic development. In the second larval instar the duct appears as a small tube surrounded by a few spindle-shaped cells. By the fourth instar a layer of matrix is present along the basal membranes of the epithelial cells, and the spindle-shaped cells form a more compact layer. Electron micrographs at this stage show that a layer of compact fibrous tissue is present which separates the differentiating fibroblasts from the epithelial cells. During the fifth instar, fibrous matrix begins to accumulate between the fibroblasts, but it is not until the first 6 days after the adult moult that most of the matrix is produced, and the cells are fully active. In the sexually mature adult the cells are smaller with less rough endoplasmic reticulum; that is, they appear quiescent.

Connective tissues also develop during the pupal stage, since the removal of the larval connective tissue is necessary to permit the reorganization of the organs to their adult form. The nervous system is no exception. In the Lepidoptera the larval neural lamella is destroyed immediately after pupation. While the perineurial cells are denuded, some migrate and form a cellular mass on the dorsal side of the abdominal connectives. As the neural lamella begins to reform, areas of matrix are produced between the cells of the dorsal mass. At this stage both the perineurial and dorsal mass cells contain large dilated cisternae of rough endoplasmic reticulum, but the Golgi complexes are very small. Matrix production ceases before the adult emerges. The cells are now much reduced in size and contain few cytoplasmic organelles. Their shape is very irregular and adjacent cells are held together by numerous desmosomes with many associated microtubules. The matrix contains typical thin, indistinctly banded fibrils.

Biosynthesis of Collagen Molecules

In common with other proteins the collagen polypeptide chains are formed on the ribosomes and then pass into the cisternae of the rough endoplasmic reticulum. At this stage the a-chains are extended at both the C- and N-terminals by about 150 residues. These terminal extensions are necessary for the proper alignment of the a-chains in the molecule. A number of post-translational modifications are made to the a-chains. These include the hydroxylation of proline and lysine, and the glycosylation of some lysine and hydroxylysine residues. The enzymes essential for these modifications are present in the cisternae of the rough endoplasmic reticulum. The collagen triple-helix is formed and the molecule is known as procollagen.

The procollagen is secreted from the cell into the extracellular matrix, where procollagen peptidases remove the C- and N-terminal extensions, and the cross-links are synthesized. Fibril formation then occurs.

During the period in which the work outlined was in progress, several investigations of the cellular pathway of collagen synthesis and secretion were made using electron autoradiographic techniques. A controversy arose as to the role of the Golgi complex in collagen synthesis, since in osteoblasts and odontoblasts there is evidence that the newly synthesized procollagen passes from the cisternae of the rough endoplasmic reticulum to the Golgi complex, and then it is secreted from the cell, whereas in chondroblasts and skin fibroblasts, it appears that some of the newly synthesized procollagen might pass straight to the extracellular matrix from the cisternae of the rough endoplasmic reticulum.

The first attempt to determine the pathway in insect fibroblasts by electron autoradiography investigated the uptake and subsequent pathway of [^{3}H]proline in the fibroblasts of the ejaculatory duct of *Locusta migratoria* and of the dorsal mass of the nerve cord of *Galleria mellonella* (Ashhurst, D. and Costin, N., 1976); [^{3}H]proline is used as a tracer since proline is found in greater amounts in collagen than in other proteins. The resulting autoradiographs were analysed using the methods of Williams, M.; these involve counting the silver grains over the cytoplasmic compartments and matrix and a stereological analysis of the relative areas occupied by each compartment on the micrographs used so that the grain counts may be expressed as the number of grains per unit area of each compartment. Both to determine grain locations and for the stereological analysis, circles, not points, are used; the diameter of the circle is determined from the data of Salpeter, *M. et al.* on the resolution of autoradiographs. For the *Locusta* and *Galleria* experiments, circles with a diameter of 4.2 mm (equal to 0.28 μm on the micrographs) were used; larger circles would have fallen over two compartments too frequently.

The results in the two species were similar; when the grain counts over the different compartments within the cells, and the extracellular matrix were expressed as the relative number of grains per unit area over each compartment, and plotted against time. If labelled molecules are passing from one compartment to another, through the cell, then the relative number of grains over compartment I should reach a maximum and start to decline, before the maximum number is reached in compartment 2, and so on. This occurs if only the rough endoplasmic reticulum and matrix are considered, and that the figures for the relative number of grains per unit area over the Golgi complex are very erratic. This is due to the very low grain counts over the Golgi compartment and to their very small size (less than 0.5 μm diameter), which means that the Golgi complex compartment is very small. It was therefore concluded by Ashhurst, D. and Costin, N. that most of the newly synthesized collagen in the locust and waxmoth fibroblasts passes directly from the cisternae of the rough endoplasmic reticulum to the matrix, and that only a minor portion passes through the Golgi complexes.

A later investigation of the pathway in fibroblasts in *Thermobia domestica* led to the suggestion that the Golgi complex is an obligatory part of the secretory pathway in these insect fibroblasts. The analysis was similar to that of Ashburst, D. and Costin, N., but circles with a diameter of almost 0.5 μm were used, and this is as large as a *Thermobia* Golgi complex. It is thus difficult to see how so many circles in the area counts fell squarely over a Golgi complex, and not over a junction between a Golgi complex and another compartment. Indeed, Ashhurst and Costin's experience was that even with smaller circles (0.28 μm diameter) very few fell over a Golgi complex only.

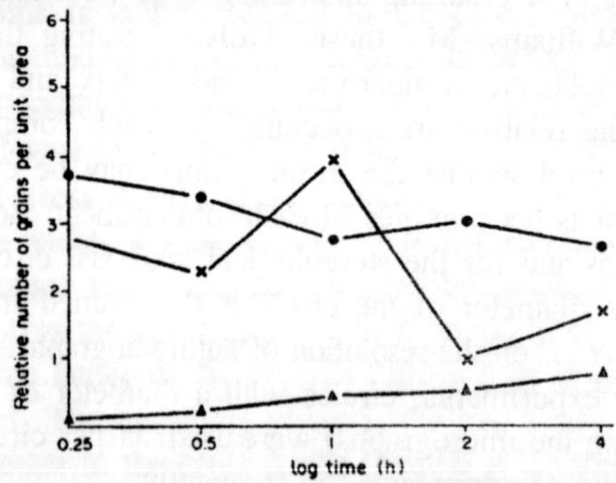

Fig. 1.4. A graph of the relative number of silver grains per unit area over the rough endoplasmic reticulum (●), Golgi complexes (×) and matrix (Δ) in locust fibroblasts and matrix at different times after a pulse of [³H]proline.

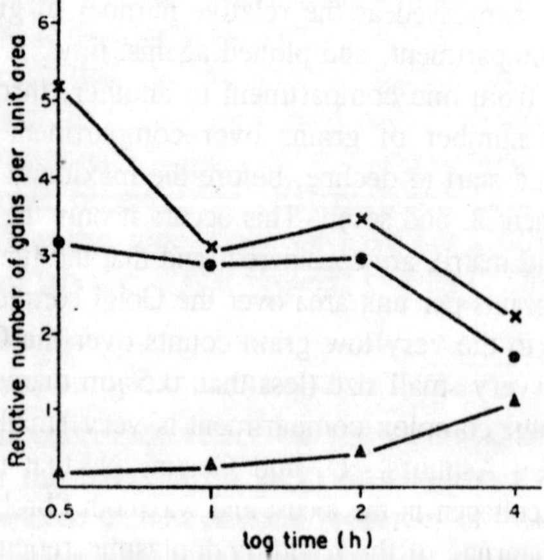

Fig. 1.5. A graph of the relative number of silver grains per unit area over the rough endoplasmic reticulum (●), Golgi complexes (×) and matrix (Δ) in Galleria *fibroblasts and matrix at different times after a pulse of [³H]proline.*

It is difficult to obtain unequivocal results with this type of study if one of the important cellular compartments is very small. The Golgi complex may be an obligatory step in the secretory pathway for at least part of the newly synthesized insect collagen. It is noteworthy that none of the enzymatic steps in the synthesis of collagen takes place in the Golgi complex. It should be remembered that these cells also produce the glycosaminoglycans and glycoproteins of thematrix, and it is known that the sulphation of chondroitin sulphate occurs in the Golgi complex.

Vitamin C

That ascorbate is an essential co-factor for the hydroxylation of both proline and lysine by their respective hydroxylases is well established. Thus a dietary intake of vitamin C is essential for animals, such as guineapigs and primates, which do not synthesize the vitamin.

No experiments have been performed to ascertain whether ascorbate is essential for collagen synthesis in insects, but it would be surprising if it is not. It is essential for the activity of prolyl hydroxylase in the shrimp, *Panaeus* sp.. Many insects do synthesize ascorbate, but it appears that many phytophagous insects cannot, and require a dietary intake for normal growth and development to proceed. These insects include many Coleoptera, Homoptera, Lepidoptera and Orthoptera. Other insects, including the cockroaches, synthesize vitamin C, or rather it is thought that it is synthesized by fat body symbiotes. It is noteworthy that Day, M. detected ascorbate in the organs of several insects including *Blatta germanica* and *Locusta migratoria*.

COLLAGEN SILKS

The existence of a silk composed of collagen was first established from X-ray diffraction patterns of the cocoon silk of *Nematus ribesii* and later in the silks of other closely related species of Symphyta. Amino acid analyses of the silk collagens show that glycine accounts for about one-third of the total residues, and the hydroxylysine level is higher than in most other collagens (about 37 per 100 residues), but the proline content is lower and hydroxyproline is absent. The periodicity of the banding pattern is estimated as 55 nm, but the intraperiod pattern cannot be clearly seen in the published micrographs.

The silks of other related species are classified as polyglycine silks, since the silk of *Phymatocera aterrima* has 660 glycines per 1000 residues. These silks are clearly worthy of further attention.

GLIAL LACUNAR SYSTEM

The glial lacunar system was first described in ganglia of *Periplaneta americana* by Wigglesworth, V.. He noticed the extensive system of extracellular spaces which in light microscope preparations can be seen between the layers of the glial cells which surround the neuropile. He suggested that the lacunae might serve as a pool for the storage of nutrients. It was reported shortly after this initial description that the glial lacunar systems of *Periplaneta,* and of a number of other species of cockroaches —*Blaberus craniifer, Blatta orientalis, Blattella germanica* and *Diploptera punctata*—contain a metachromatic material, labile to hyaluronidase digestion, which was identified biochemically as hyaluronate by Ashhurst, D. and Patel, N..

Since these initial findings in the cockroaches, it has been demonstrated that other insects, and also other arthropods, do possess a glial lacunar system, but that it is not so extensive as in cockroaches, where it occupies about 15-18% of the volume of the meso and metathoracic ganglia in *Periplaneta* adults.

The glial lacunar system is clearly demonstrated in preparations of ganglia stained with alcian blue at pH 2.5. In the locust, *Locusta migratoria*, and the stick insect, *Carausius morosus*, it is much less extensive than in *Periplaneta,* but histochemical studies of these three species using alcian blue and other histochemical techniques confirm that the glial lacunar system contains only hyaluronate. Martoja, R. and Cantacuzene, A. thought that the glial lacunar system of *Locusta* is PAS-positive, and they suggested that it contains glycoproteins in addition to hyaluronate. Close examination of sections of locust ganglia on which both the PAS test and alcian blue staining at pH 2.5 have been performed indicates that the alcianophil areas are distinct from the PAS-positive areas, which are in fact the glial cells with much glycoprotein (not glycogen) in the cytoplasm.

The glial lacunar system looks empty in electron micrographs apart from some fine amorphous material, or fibre-like concentrations of it. The better preservation of the tissue enables the system to be traced into the neuropile where narrow extracellular channels ramify

TABLE 1.3 : HISTOCHEMICAL REACTIONS OF THE GLIAL LACUNAR SYSTEMS OF THE COCKROACH, LOCUST AND STICK INSECT THORACIC GANGLIA

	Glial Cock-kroach	*lacunar Locusts*	*system Stic insect*
Alcian blue, pH 2.5	4+	3+	1-3+
Alcian blue, pH 1.0	0-1+	0	0
Alcian blue, pH 5.7 + 0.05 M $MgCl_2$	4+	4+	3+

Alcian blue, pH 5.7 + 0.1 M $MgCl_2$	4+	3+	2+
Alcian blue, pH 5.7 + 0.2 M $MgCl_2$	0	0	0
Alcian blue, pH 5.7 + 0.4 M $MgCl_2$	0	0	0
Alcian blue, pH 5.7 + 0.6 M $MgCl_2$	0	0	0
Alcian blue, pH 5.7 + 0.8 M $MgCl_2$	0	0	0
Alcian blue, pH 5.7 + 1.0 M $MgCl_2$	0	0	0
PAS	0	0	0
HID	0	0	0
Hyaluronidase-AB pH 2.5	1–2+	2+	0–1+
Hyaluronidase control-AB pH 2.5	4+	3+	1–3+
Neuraminidase-AB pH 2.5	4+	3+	1–3+
Neuraminidase control-AB pH 2.5	4+	3+	1–3+

between the axons; indications of these channels may actually be observed in alcian blue preparations. It is noteworthy that although alcian blue staining fails to reveal a glial lacunar system in lepidopteran ganglia, narrow lacunae are present among the axons in all published electron micrographs.

The inclusion of the glial lacunar system among the connective tissue matrices is justified because the lacunae are extracellular and the substance within them is hyaluronate, a typical connective tissue molecule.

Development of the Glial Lacunar System

It was noted by Martoja, R. and Cantacuzene, A. that the glial lacunar system can be identified from the end of the first larval instar in *Locusta* and *Gryllus*. A recent examination of developing mesoand metathoracic ganglia of *Locusta* and *Periplaneta* has shown that the glial lacunar system is present in *Periplaneta* at hatching, but not until the second instar in *Locusta*. The variation with the results mentioned above may be because the first instar locusts used by Ashhurst were not at the end of the instar. From the earliest stages the histochemical reactions of the material in the lacunae are the same as those in the adult. That is, hyaluronate is present in the glial lacunar system throughout the life of the insect.

The volume of the glial lacunar system of the meso- and metathoracic ganglia of *Periplaneta* increased steadily from about 1 % in the newly hatched larva to between 15 and 18% of the total volume of the thoracic ganglia in the adult.

Function of the Glial Lacunar System

The mechanism of nerve conduction in insects poses many questions since the ionic concentrations in the haemolymph would affect the conduction of potentials adversely if a selective barrier, located in the glial cells, did not separate the haemolymph and axons. The ionic composition of the fluid bathing the axons is controlled by ion pumps in the glial cells. The extracellular clefts around each axon are very narrow, and it is suggested that the hyaluronate in the glial lacunar system acts as an extracellular cation reservoir, particularly for sodium ions. Varying concentrations of potassium, calcium, magnesium and phosphate ions have been recorded in the glial lacunar systemof *Locusta migratoria*. Thus, the glial lacunar system with its content of hyaluronate appears to be essential for the proper functioning of the insect nervous system.

ELASTIC FIBRES

A second type of connective tissue fibre was described by Baccetti, B. and Bigliardi, E. in the wall of the dorsal vessel of the orthopteran *Aiolopus strepens*. Since the fibres appeared to be labile to elastase digestion, they were identified as elastic fibres. Later, Locke, M. and Huie, P. found similar fibre bundles in the neural lamella and pericardial connective tissue matrix of *Calpodes ethtius* larvae, and Crossley, A. saw them in "connecting ligaments" between the pericardial cells of *Calliphora erythrocephala*. They also occur in the connective tissue matrix around the mesenteron of *Periptaneta americana*.

The fibre bundles are not easily distinguished in sections of material prepared for electron microscopy by the usual procedures. Locke, M. and Huie, P. found that they are very obvious when a peroxidase reaction is performed on the tissue, when sections of tissue fixed in glutaraldehyde only are stained with phosphotungstic acid, or when tannic acid is added to the glutaraldehyde fixative. Using these methods they were able to distinguish two types of fibril in the fibre bundles: thick fibrils, about 40 nm in diameter, are separated by a 70 nm space in which microfibrils, about 6 nm in diameter, are found.

The identification of these fibre bundles as a type of elastic fibre is based entirely on the known elasticity of the tissues. The presence of these fibre bundles in the neural lamella of the abdominal nerve cord of the adult female locust confirmed this hypothesis because the abdomen stretches to about three times its normal length during oviposition. It appears doubtful, however, whether these fibrils should be termed elastic fibres, as this nomenclature implies that they are homologous to vertebrate elastic fibres.

Vertebrate elastic fibres have an amorphous, electron-lucent core, composed of the protein elastin, which is surrounded by microfilaments. Despite an extensive study of invertebrates from different phyla, Sage, H. and Gray, W. were unable to find either the protein elastin, or fibres which stain with the standard methods used for vertebrate elastic fibres. Thus, they conclude that elastin is found exclusively in vertebrates. The fibres with elastic properties in the invertebrates are not, therefore, homologous to vertebrate elastic fibres. Thus the composition of the fibres described in insects, and of those of other invertebrates identified as "elastic fibres" by their affinity for the dye spirit blue, has yet to be established. The lability of the fibres of *Aiolopus* to elastase digestion, mentioned above, is probably of little significance, since commerical samples of connective tissue enzymes all contain large amounts of nonspecific proteolytic enzymes.

CONNECTIVE TISSUE ENZYMES

There is, as yet, no information about the degradative enzymes in the connective tissues; some must be present, as the connective tissues have to be modified during growth and development. There are, however, a few reports of connective tissue enzymes which are produced to destroy the connective tissues of other animals; these enzymes must be very commonly produced by insects.

A collagenase has been isolated from first-instar larvae of *Hypoderma lineatum*. The larvae of *Hypoderma* are endoparasites of cattle, and after penetrating the skin they migrate through the connective tissues, aided by the collagenase produced by the salivary glands. The enzyme has been characterized. It cleaves the helical region of the collagen molecule and has a molecular weight of 24,000. The amino acid composition suggests that it is similar to the serine proteases, as is the collagenase found in the fiddler crab, *Uca pugilator*. It differs from the vertebrate collagenases, which are metalloenzymes, since it is not inhibited by EDTA. It is suggested by Lecroisey, *A. et al.* that the insect and crab enzymes which function as digestive enzymes rather than tissue enzymes facilitating morphogenesis, may be serine proteases, related to trypsin.

Hyaluronidases are also produced by insects. An extract of the testis of *Drosophila melanogaster* was found to reduce the viscosity of hyaluronate, though its enzymatic nature was not conclusively confirmed. Hyaluronidases have also been isolated from the testis and mucus accessory glands of the male bee, *Apis mellifera,* and the hornet, *Vespa orientalis*. The enzymes from both sources are similar; they have a pH optimum of about 4.5, they degrade hyaluronate and chondroitin-6-sulphate, and in contrast to mammalian testicular hyaluronidase, they are not inactivated by a temperature of 56° for 3 h. It is suggested that they serve to remove mucus from the female reproductive tract. More recently, the hyaluronidase in bee venom with a molecular weight of 60,000 was purified. Since the purpose of the research was to obtain pure samples of the enzyme for work on the human allergens in bee venom, it was not characterized.

Breakdown of Connective Tissue

While some modification of the connective tissues must occur during growth and development in all insects, the most extensive breakdown of connective tissue matrices occurs during the metamorphosis of the endopterygote insects. The disintegration and removal of the neural lamella around the nervous system of Lepidoptera at the beginning of the pupal stage is well documented, and it appears to be an essential prerequisite for the shortening and reorganization of the larval nerve cord to the adult form. Adipohaemocytes (granular haemocytes) are in close proximity to the neural lamella at this time, and while their role in phagocytosing the remnants of the connective tissue is establised, there is as yet no evidence that they release lytic enzymes. Acid phosphatase is present in the granules of the adipohaemocytes of *Galleria* (Ashhurst, unpublished observation) and thus if the granules contain this lysosomal enzyme, it is very probable that they also contain other hydrolytic enzymes, which could be released at the onset of metamorphosis. A similar process occurs in the Diptera during metamorphosis; Whitten, J. observed the lysis of the connective tissues in dissections of *Sarcophaga* pupae and the removal of the fragments by granular haemocytes.

The hormonal control of the changes in the nerve cord of *Galleria mellonella* was investigated by Pipa, R.. He found by transplanting nerve cords from pupae at known ages into larvae, pupae and adults, that the process of axon shortening and connective tissue removal is initiated during a short critical period in the early pupa. It has been demonstrated in a series of experiments on *Galleria* in which the brain and associated glands were removed from pupae of different ages that the brain, and to a lesser extent the corpora cardiaca and corpora allata, have a role in the control of axon shortening; the evidence suggests that the brain may act by producing a prothoracotropic hormone. In brainless pupae, shortening, and the associated connective tissue removal and later re-formation, are induced by injection of synthetic α-ecdysone.

FUNCTIONS OF CONNECTIVE TISSUES

Connective tissues are supportive. They bind cells together, hold organs in position and in many animals form the main skeletal tissue. It is only relatively recently that their wider significance has been appreciated, as information about the properties of the glycosaminoglycans has accummulated. The glycosaminoglycan-protein complexes, or proteoglycans, are very large and have a high density of anionic charges along the length of the glycosaminoglycan chains. As a result the molecules are very highly expanded *in vivo* and occupy a significant proportion of the volume of the matrix. They form an anionically charged meshwork which appears to act as a sieve affecting the movement of ions and molecules within the matrix; anionically charged ions and molecules may be repelled, cationically charged ions may be hindered while crossing the matrix, or excluded from it, by virtue of their size and shape. Chondroitin sulphates also appear to effect the formation of collagen fibrils.

The mechanical strength of a connective tissue matrix will depend on the density and orientation of the collagen fibrils within it, and those with nonfibrous forms of collagen will be less strong. The presence of proteoglycans in matrices means that they form a potentially selective barrier, and in all animals, exchange between the blood and tissues occurs across a layer of connective tissue. It has been emphasized throughout this chapter that all insect organs are covered by a connective tissue matrix which effectively separates the cells from the haemolymph. The role of the neural lamella has been extensively investigated, and while Eldefrawi, M. and O'Brien, R. found evidence for a mechanism discriminating against large size and polarity, which slows the passage of molecules through the neural lamella, the actual barrier to the free passage of ions and molecules is located in the perineurial and glial cells. It appears, however, that connective tissue might form the barrier which protects muscle fibres from adverse concentrations of glutamate in the haemolymph. The ion-binding properties of hyaluronate have been implicated in the formation of a cation reservoir in the glial lacunar system.

It was suggested by Whitten, J. that the thin connective tissue strands observed in many insects might form channels for the passage of hormones and other substances between organs. This idea has not been substantiated by other workers.

More recently it has become apparent that connective tissue matrices are essential for morphogenetic movements and cell differentiation. Basement membranes are found in vertebrate embryos prior to gastrulation and sulphated glycosaminoglycans must be present in the basement membrane of developing salivary glands for normal acini to be formed. In a study of limb bud development, Toole, B. found that the presumptive chondroblasts move into position in a matrix of hyaluronate, but this must be removed by hyaluronidase before differentiation and the secretion of the cartilage matrix can occur. These are some examples of early work in this area, which has expanded rapidly to confirm the intimate association of the connective tissues in developmental processes. While such studies on insects have not commenced, it is significant that Hoglund, L. in his biochemical study found glycosaminoglycans in *Calliphora* embryos from late cleavage onwards.

CONCLUSIONS

Research on insect connective tissues over the last decade has shown that the chemistry of the connective tissue matrices of insects is essentially similar to that of other animals. Perhaps the most significant finding is that the fibrous collagen molecule is of the same type as that of many other invertebrates and vertebrates, namely the type I trimer. The next step is to characterize the glycosaminoglycans and glycoproteins in more detail biochemically, since histochemical identification has inherent limitations

The greatest problem to be overcome in any biochemical work is the small amount of material available from any known site, since the characterization of an extracted substance should include knowledge of its precise location in the insect. The newly developed immunocytochemical techniques offer a means by which small quantities of substances can be located within a tissue, and it is obvious that these could give us a great deal of information about the components of insect connective tissue matrices. Their widespread use will, however, depend upon the cross-reactivity of antibodies made to collagen and the other macromolecules from vertebrate tissues with the insect antigens, since it is not easy to extract and purify sufficient antigen from insects to immunize a rabbit, or other mammal. While monoclonal antibodies have been made to some components of the *Drosophila* matrix, it is not yet known if these will cross-react with the molecules of other species of Diptera, or those of other insect orders. To produce antibodies to the connective tissue molecules of many different species of insect would be laborious indeed. Specific, cross-reacting antibodies made to the collagens, proteoglycans and glycoproteins of avian and mammalian species will offer the most economical way to locate these macromolecules in the various insect tissues. These new techniques will provide many new possibilities for the development of insect connective tissue research. Throughout this chapter I have tried to stress the similarities between the connective tissues of insects and other animals. While insects do have some peculiar connective tissue matrices, such as the layers found under some midgut epithelia, in general, insect connective tissues are similar in their morphology, biochemistry and development to the connective tissues of other animals.

2

Body Covering

The bones, cartilage and tendons of the vertebrate principally serve an endoskeletal function. The integument of insects, on the other hand, serves so many more purposes: it is the interface between a living animal and an environment from which both beneficial and harmful influences come. Thus, besides serving as an exoskeleton, the integument is a rich sensory depot which must perceive stimuli in many different forms.

The integument is very far from being an inert substance. While it consists of a cuticle and an epidermis, neither of these layers has any physiological integrity without the other. The integument is a collection and an historical record of synthetic activities performed by the animal. But this record is not static as a rune, the kinds and qualities of layers come and go, especially during metamorphosis. It is hardly surprising that, with great structural changes occurring from instar to instar, equally great physical and chemical changes also occur. It is the purpose of this chapter to document some of the striking and important properties of the integument and to make inferences as to their significance in the lives of "... these narrow Engines ...".

EPICUTICLE

General Properties

The epicuticle is an extraordinarily complex and very thin composite structure that overlies the chitin-bearing procuticle of insects. An epicuticle generally covers the entire external surface of an insect but there are a few exceptions associated with special functions, the most notable being the surfaces of some chemoreceptive sensillae first described by Slifer, E.. While the fore- and hindguts of insects are also covered with an epicuticle, the midgut, which is endodermal in origin, lacks an epicuticular covering.

The functions of the epicuticle, shown experimentally or inferred, are many and varied and in most cases it is not yet clear how precisely function is related to structure. The epicuticle makes important contributions to water homeostasis both in the uptake of water through the integument of terrestrial insects such as the flea *Xenopsylla brasiliensis* or in the aquatic alderfly, *Sialis lutaria,* as well as through resistance to water loss first studied in any experimental detail in the cockroach *Periplaneta americana* by Ramsay, J.. The epicuticle is thought to set limits on the dimensions an exoskeleton can attain during intrastadial growth of holometabolous larvae such as has been documented for the skipper, *Calpodes ethlius*, and the silkworms *Bombyx mori*. Similarly, the extent to which a recently ecdysed cuticle may expand, or that of a sap- or blood-sucking insect may distend, during engorgement are thought to reside in the mechanical properties of the epicuticle in each particular case.

Additional functions for the epicuticle have been inferred, but the structural bases for such functions have remained completely elusive to date. For example, the epicuticle must be selectively permeable to the transport of moulting fluid precursors into the apolysial space (necessary for the partial digestion of the outer-lying presumptive exuvial cuticle) and to the return of the endocuticular digests into the body proper of the pharate instar. At the same time, the pharate epicuticle must itself be protected from chemical dissolution by moulting fluid enzymes.

Although we have no direct quantitative measurements of the mechanical properties of the epicuticle it has been inferred from its bending and fracture characteristics that it is in some ways analogous to a lacquer, stronger in compression than in tension. The epicuticle may also serve as a chemical reservoir, either as a depot for the storage of metabolic waste products or for the brief retention of defensive secretions and possibly even for juvenile hormone.

Turning to the structure of the epicuticle, this subject is made all the more difficult because of arguments over homology that have been in progress for about 150 years. The nomenclatural problems remind us of Alice: "... What's the point of their having names if they don't come when you call them?" Nonetheless, a generalized view of the epicuticle has been developing over the last several years and can be summarized as follows. It consists of four principal layers each of which is of unique composition and properties. There is an outermost cement layer that is subtended by a wax layer. Below the wax layer there is an outer epicuticle, and below that an inner epicuticle. Unlike the layered subdivisions of the chitin-bearing procuticle, the layers of the epicuticle are synthesized in a temporal sequence that is the reverse of that in which they are anatomically situated in a fully differentiated cuticle. The inner and outer epicuticles are formed before apolysis and thus define the reality of the pharate cuticle while the cement and wax layers follow apolysis.

Cement Layer

The cement layer is generally regarded as the result of the combined secretory products of specialized dermal glands (specialized epidermal cells) and thought to consist of proteins and lipids that have been stabilized by various polyphenolic substances. The distribution and relative thickness of the cement layer varies enormously in the

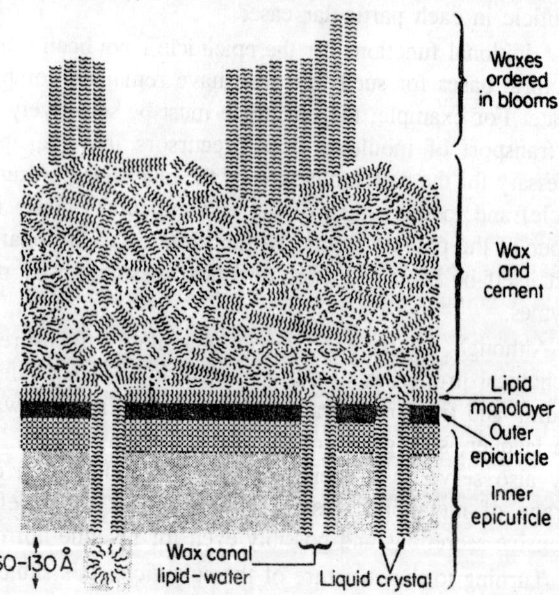

Fig. 2.1. Schematic diagram of an insect epicuticle. In this instance the most complex situation known is illustrated. Thus, wax is associated with the cement layer and wax also occurs in a bloom above the cement layer.

same or in different insects. For example, most of the exoskeleton of *Rhodnius prolixus* is covered by a cement layer, but certain regions of the abdominal tergites actually lack a cement layer and some caterpillars are said to be only partially covered with a cement layer as in *Diataraxia oleracea*. The cement layer may be completely absent as from the honeybee.

It has very often been suggested that the properties of the cement layer are roughly analogous to those of commercial shellac, the latter being the secretory products of the lac insect, *Laccifer lacca*. Its formation has been compared with that of the cockroach ootheca (as, incidentally, has the tanning of procuticle been compared with oothecal tanning) which is secreted by separate collaterial glands and the combined products of secretion after mixing.

The formation of the cement layer has been principally studied by Wigglesworth who has recently summarized its morphogenesis and properties in the sucking bug *Rhodnius prolixus*. On ecdysis, the new teneral cuticle is hydrophobic but shortly after eclosion, secretions from the specialized dermal glands either penetrate or cover the cement layer so that it becomes hydrophilic. Some hours later the same structure becomes hydrophobic again, possibly because it has become stabilized chemically through processes analogous to tanning and/or because the cement layer itself becomes impregnated with waxy material which is hydrophobic in nature. The cement layer is thought to function primarily by protecting the subtending wax layer against possible abrasive or impact damage. A comparative study of cuticles with and without cement layers would contribute much to our very limited knowledge of this epicuticular layer.

Wax Layer

The wax layer is situated just below the cement layer in insects where the latter is in fact present. The composition of insect waxes has been investigated over the last few decades, the earlier work being summarized by Gilmour, D. and that of the interim by Jackson, L. and Blomquist, G.. Generalizations about cuticular waxes are fraught with difficulty because of variations in solvent extraction procedures and conditions. Indeed, composition has recently been shown to vary biologically also, and variations in lipid class occur between individuals of the same species, as well as at different times of development .

In general, the composition of insect cuticular waxes is dominated by hydrocarbon molecules varying in length from 12 to 31 carbons and by esters of even-numbered fatty acids and alcohols. The physiological

TABLE 2.1 : MAJOR CLASSES OF INSECT CUTICULAR LIPIDS

Fraction	*Pteronarcys californica*		*Anabrus simplex*	*Periplaneta australasie, Periplaneta brunnea, Periplaneta fulginosa*	*Lucilia cuprina*		*Cynthia ricini*	*Tenebrio Molitor*	*Melanoplus Sanguinipes*
	Larva	Adult			Pupa	Adult			
Hydrocarbon	3	12	48-58	90	33	62	5	10	60
Wax esters	1	4	9-11				16	13	28
Triglyceride	78	7	—	7	25	16	—	—	1
Free fatty acids	12	49	15-18	2	16	7	Tr	5	6
Aliphatic alcohols	—	—	2- 3	—	—	—	70	58	2
Sterols	1	18	—	1	2	—	Tr	1	1
Others	5	10	15-18	—	15	24	8	13	—

significance of these compounds may well lie in their physical attributes: saturated and esterified fatty acids are associated with hardness in commercial waxes while unsaturated analogues tend to be soft and even greasy. The waxes are mainly long-chain saturated alcohols esterified with acids. However, owing to the extremely labile nature of lipids, the precise composition of epicuticular wax for any insect must be regarded as tentative.

The structural arrangement of wax in the cuticle is not a simple matter, nor can we describe a discrete wax layer. At least four different variations of arrangement have been described to date. A lipid monolayer was described by Beament, J. and this layer can be seen immediately above the outer epicuticle. A more complex arrangement based on studies of several species has been visualized by Locke, M. who shows that in some insects wax may also be combined with the cement layer as well as extending above the surface of the cement layer in the form of wax blooms commonly associated with Heteroptera and Homoptera and also very many beetles. Finally a structural role for lipid in the procuticle has been intimated by Wigglesworth, V..

The lipids of the wax layer are probably secreted by the epidermis just prior to ecdysis. Generally, when lipid is initially released at the surface of an insect it is in a mobile liquid phase but is transformed into wax at ecdysis. They are then transported through the pore canal system to the somewhat finer and arborescing wax canals. Locke, M. has concluded that the final wax product is actually synthesized in the wax canal filament where esterases modify final wax precursors that have come from the epidermis. His conclusions were based on studies of several insects such as the honeybee, waxmoth, the skipper *Calpodes ethlius* and the beetle *Tenebrio molitor.* Against this, it has been claimed that there is a wax layer on the caterpillar *Diataraxia* which is said to lack pore canals, and in *Calpodes ethlius*. Clearly it would be of interest to reinvestigate many of these allegedly unusual cases with the considerably more sophisticated techniques presently available.

The physiological significance of the insect wax layer has been best documented in terms of its relationship to waterproofing. The relationship between cuticular transpiration and temperature change has been studied in some detail in the case of the butterfly *Pieris*. For this animal there is an initial critical transition temperature at which there is a sudden increase in cuticular water permeability followed by a plateau region of little change and again another transition temperature at which permeability further increases.

At present it remains uncertain whether or not the observed changes in cuticular permeability are due to either failure or changes in the lipid phase that might be related to lipid melting points, or whether there might be shifts in structure toward liquid crystal phases as suggested by Beament, J.. Recently, a number of branched lipids have been reported from several different desert beetles, *Eleodes, Cryptoglossa* and *Centrioptera,* but it is not clear how these compounds relate to permeability vis-a-vis lipid structure.

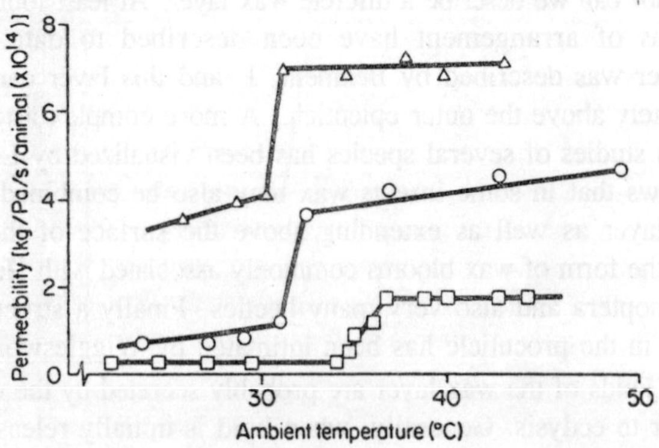

Fig. 2.2. Cuticular water permeability as related to changes in temperature. Squares denote pupae of Tenebrio molitor, circles, the larvae of Rhodnius prolixus, the triangles, pupae of the butterfly Pieris rapae.

Numerous biological observations as to changes in wax layer properties have been recorded in recent times. For example, diapausing pupae of the tomato sphinx, *Manduca,* produce three times more epicuticular wax than do non-diapausing pupae, and the inference is that changes in epicuticular wax synthesis could be mediated by neuroendocrine events.

Outer Epicuticle

The outer epicuticle (often designated as cuticulin) appears thus far to be of universal occurrence as an epicuticular layer of the exoskeleton, as well as occurring in the linings of tracheae and serving as the insertion layer for the tonofibrillae of muscles. It has been described by Locke, M. as a trilaminar layer of about 12-18 nm in thickness. Its chemistry and composition remain unknown but its resistance to degradation suggests properties similar to polythene-like polymers and to quinone-tanned protein.

The outer epicuticle is the first-formed layer of the pharate cuticle. It is derived from specialized regions of the apical membrane of the epidermal cells, namely the plasma membrane plaques. During development discrete plates appear above the membrane and gradually form a continuous layer through accretion. Shortly after the outer epicuticle has been formed, small pores of about 3 nm diameter can be detected in this layer and it has been suggested that these pores allow the egress of apolysial droplets into the apolysial space for the partial digestion of the presumptive exuvial cuticle and for the transport of the products of digestion into the body of the pharate instar.

In view of the critically important functional roles commonly ascribed to the outer epicuticle, whatever it may be called, it is surprising that the composition of this material has so seldom come under the scrutiny of those of chemical bent. Because of its position in the cuticle, it has important contributions as a selective permeability barrier in the metabolism of old cuticle, particularly during the course of Pharate cuticle development. The outer epicuticle apparently limits the extent to which intrastadial growth, through stretching or intussusception can occur. Similarly, it is thought to define the final dimensions to which the exoskeleton will ultimately expand following an ecdysis, and thus give a basis for the observations variously known as "Brooks' rule", "Dyar's rule" or "Przibram's law".

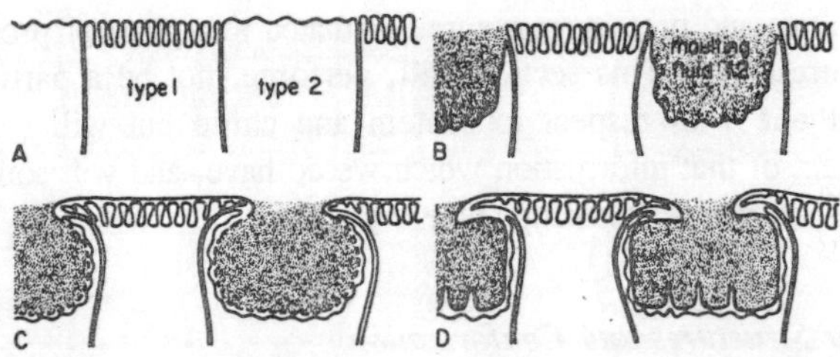

Fig. 2.3. Diagram summarizing formation of the outer epicuticle in the evaporative surface area of the stink bug, Nezara viridula. *Stage A represents differentiation of epidermal cells prior to secretion of apolysial fluid. The plasma membranes of the two cells are continuous. B shows the apolysial fluid associated with a type-2 cell. The forming outer epicuticle occurs in small patches above the microvilli. In C, outer epicuticle formation proceeds at a greater rate over type-1 cells laterally displacing the forming cuticle.*

The intricate and often complex surface patterns of insects, specialized areas such as diffraction gratings, plastrons, strigils and the like, are all formed at the same time as the outer epicuticle, so that its morphogenetic significance is thought to be great. In this respect, the "chicken or the egg" question arises: to what extent is the outer epicuticle itself the determinant of final exoskeleton shape *vis-a-vis* the crenulated microvilli of the apical plasma membrane plaques which actually secrete the products of the outer epicuticle.

Inner Epicuticle

The inner epicuticle is an optically isotropic layer thought to consist of polymerized lipoprotein stabilized by quinones, a conclusion reached in early studies by Dennell, R.. It is thought to be a product of secretion of the epidermis to which contributions from the oenocytes have been added. It has also been postulated that the inner epicuticle serves as the reservoir for extracellular enzymes associated with wound repair.

PROCUTICLE

General Properties

The procuticle of the insect integument has been classically understood, and without exception, to be the chitin-bearing part of the cuticle proper. While the chitin-free epicuticle has been the subject of several interesting controversies, particularly with reference to permeability properties, differing interpretations of various aspects of the procuticle have led to most intriguing arguments of a fundamental nature. Studies of the principal components of procuticle, chitin and protein have not developed apace, and we know considerably more about the molecular dimensions and configurations of chitin than we do of protein. Even very fundamental problems, such as how exactly are the chitins and proteins structurally situated in a piece of procuticle, remain unresolved. This section will, perforce, not be a particularly integrated one with respect to protein and chitin but will trace the development of that information which we do have, and will sometimes suggest functional relationships where possible.

Chitins

Molecular Structures and Conformation

Chitin, poly-β-(I → 4)-N-acetyl-n-glucosamine, is an abundant and widespread structural polymer akin to the celluloses and is synthesized by fungi, protozoans and most of the protostomian invertebrates. In insects it occurs naturally in crystalline form but it is crystallographically heterogeneous, and three polymorphs which differ both in the packing and polarities of adjacent chains have thus far been described.

α-Chitin is the most prevalent of the chitin polymorphs and a detailed structure for it was proposed by Carlstrom, D.. He confirmed the orthorhombic unit cell of earlier crystallographers, defined its dimensions and space group symmetry (adjacent chains are anti-parallel) and showed that the crystallites arise from hydrogen-bonded sheets of chains. The Carlstrom structure has been the subject of much refinement over the last 20 years and a recent treatment by Blackwell's group appears to have overcome some of the "noise" problems inherent in the 1957 model.

Thus, α-chitin is currently understood to consist of an orthorhombic unit cell with the following dimensions: 4.74Å in the crystallographic *a*-axis, 18.86 A in the *b*-axis and 10.32 Å in the *c*-, or fibre, axis. The unit cell consists of disaccharide sections of two anti-parallel chains bonded in two ways. There is an intramolecular hydrogen bond (03-H ... O5) and successive chains are connected by C = O ... H — N bonds, thus forming sheets along the *a*-axis. The CH_2OH side-chains adopt very different conformations on each side of the two chains within the unit cell structure: the CH_2OH group on one side forms an intramolecular hydrogen bond with the carbonyl group of the next residue (O6–H ... 07'); the CH_2OH group of the other chain forms an intermolecular or intersheet bond (O6–H ... O7). Different amide groups also participate in hydrogen bonding. In Blackwell's refinement there is a statistical mixture of the CH_2OH bonding groups. Finally, the basic α-chitin chain is sterically rigid except for rotational freedom about the side-chains and this facilitates transconformational changes that can be induced chemically and mechanically.

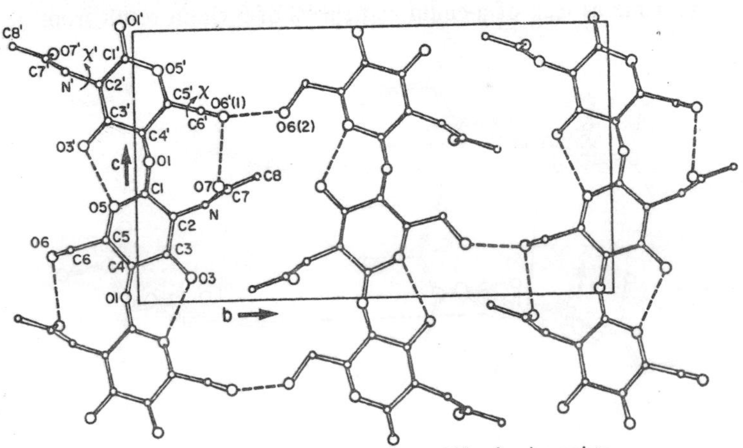

Fig. 2.4. Proposed structure for a-chitin showing points of rotation (Z' and x) about the side-chains.

β-Chitin was first distinguished from α-chitin by Lotmar, W. and Picken, L. in diffraction studies of the squid pen *(Loligo)*. It must be considered as rare among insects because it has thus far been identified only in the cocoons of the figwort weevils, *Cionus* and *Cleopus,* and that of the cerylonid beetle, *Murmidius*. The structure of β-chitin has also been the subject of refinement studies through the years and has recently been characterized in some detail.

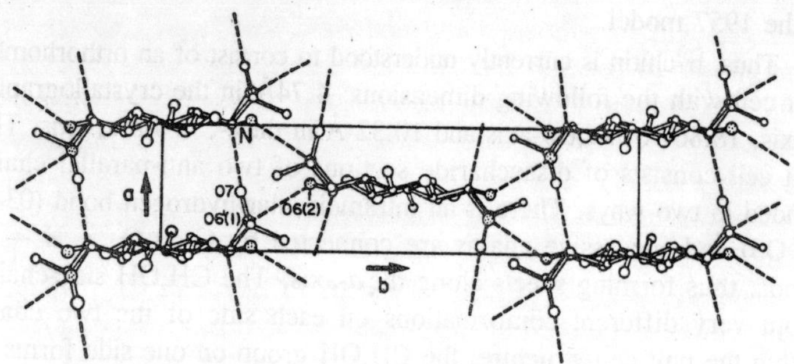

Fig. 2.5. ab *projection of α-chitin showing the C7=07 ... H-N and the 06(2)-H ... 06' intermolecular bonds.*

β-Chitin consists of a monoclinic unit cell with the following dimensions: 4.85 A in the crystallographic *a*-axis, 9.26 Å in the b-axis and 10.38 Å in the *c*- or fibre axis. The unit cell thus contains but a single repeating disaccharide (the β-axial dimension of fl-chitin being only about one-half that of α-chitin). The chains are arranged in parallel so that the sense is opposite that of achktka. The char conforma ~ xon \s tire same as teak of a-chitin and sheets of β-chitin result from

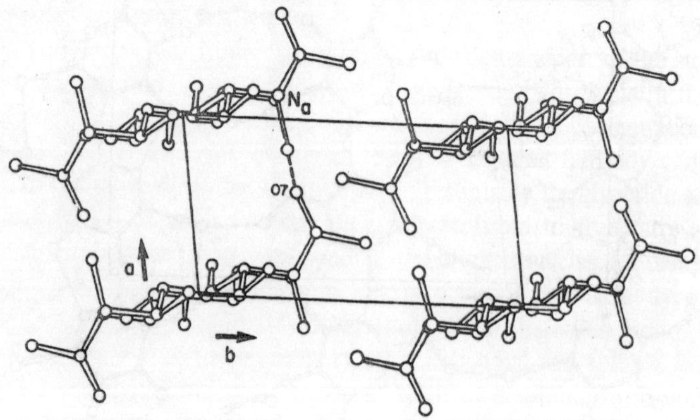

Fig. 2.6. Proposed structure for anhydrous β-chitin in ab projection showing the C7 = 07 ... H—N intermolecular bond.

C=O... H—N hydrogen bonds. CH_2OH groups are hydrogen bonded to the carbonyl group of the next chain along the *a*-axis. There is no intermolecular hydrogen bonding between sheets, and this provides a basis for the ease of solvation in this polymorph.

γ-Chitin was discovered and partially characterized by the indefatigable crystallographer Rudall, K. in studies of squid *(Loligo)* stomach. In insects this polymorph is known from the cocoons of the weevils *Prionomerus* and *Rhynchaenus,* the leaf beetle *Donacia,* the spider beetles *Ptinus* and *Gibbium* and the great dung beetle *Pacylomera* as well as from the peritrophic membranes of the cockroach *Blaberus, Schistocerca,* the caterpillar *Antheraea pernyi* and the sawfly *Phymatocera.* It is of interest to note that these same membranes in honeybees and other aculeate Hymenoptera consist only of a-chitin.

The unit cell dimensions of γ-chitin are regarded as tentative (awaiting additional confirmatory data) and are suggested to be as follows: ~ 4.7 Å in the crystallographic a-axis, ~ 29 Å in the β-axis and ~ 10.3 Å in the *c*- or fibre axis. It will be noted that the β-axial dimension of γ-chitin is about three times that of β-chitin and this leads to the suggestion that the γ-chitin molecule folds back on itself. γ-Chitin is similar to β-chitin in its propensity for hydration and solvation. The ease with which both β- and γ-chitin form hydrates mitigates against their usefulness as skeletal fibres. The fact that a-chitin is considerably less susceptible to hydration effects may also be *one* of the reasons why most insects must moult in order to appreciably increase their surface dimensions.

All three polymorphs of chitin, though variable in degree of crystallinity, occur in insects as microfibrils and are classified as viscoelastic polymers. That is, when macroscopic samples of all of these chitins are deformed by tensile loading their mechanical properties such as stiffness, strength and extensibility vary within and among polymorphs with differing rates of deformation. Precise interpretation of the observed mechanical behaviour is made difficult because of differences in chitin samples with respect to crystallinity, orientation and microfibrillar packing in actually testable macrofibres.

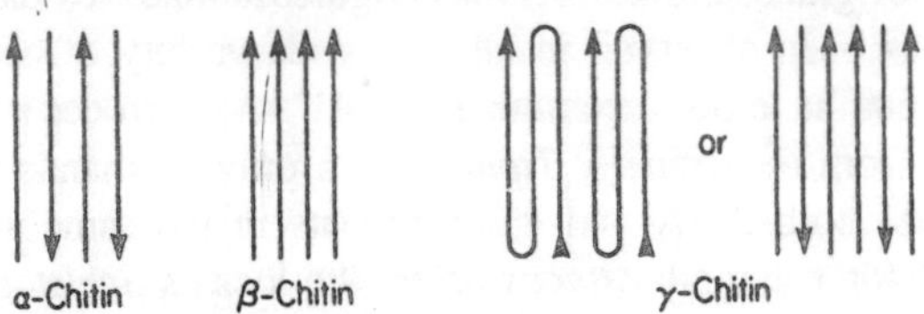

Fig. 2.7. Schematic representation of chitin chains in the unit cells of α-, β- and γ-chitin.

Nonetheless, mechanochemical studies of chitin polymorphs indicate some important properties that ultimately derive from the structure of the crystallites concerned. Owing to the inter-chain bonding characteristics of α-chitin *vis-a-vis* β- and γ-chitin, γ-chitin appears to deform viscoelastically by neighbouring chains slipping past one another while the deformation characteristics of β- and γ-chitin are consistent with molecular conformational changes. While the tertiary structure of γ-chitin remains equivocal, mechanochemical evidence is consistent with a ta-shaped unit cell rather than the TIT-shaped cell. One can envisage the native conformation of γ-chitin as crudely analogous to the cross-β conformation described from the eggstalk silks of *Chrysopa.*

Chitin crystals always occur in insects as collections of microfibrils, and measurements from many different preparations show that these microfibrils range from about 25 Å to 30 Å in diameter. Microfibrils are in turn always associated with protein in a chitin-protein complex, be it cuticle, cocoon or peritrophic membrane. However, the insect exoskeleton is thought to contain only complexed a-chitin in microfibrils consisting on average of between 18 and 21 chains passing through a crystallite at any particular level, implying 3 sheets containing 6 or 7 chains.

The structure of chitin has been classically defined on the basis of diffraction data obtained after chemical purification of a chitin-bearing sample, hence the "pure chitin lattice". However, it has long been known that X-ray diffraction patterns of intact cuticles are "modified" owing to the presence of protein. Rudall has extended these observations over the years and has defined two basic chitin-protein complexes: a 31 Å axial repeat system in which diffraction spacings are associated with three chitobiose molecules (or 6 GlcNAc residues) and a 41 Å axial repeat system based on four chitobiose (8 GlcNAc, residues).

The origins and functional significance of these two repeating systems are not at all clear. It could be that in the 31 Å system, for example, every sixth residue differs from the preceding ones, a substitution of glucosamine for N-acetylglucosamine. To date, the 31 Å system has been observed in all of a wide variety of bees, wasps and beetles that have been examined; the 41 Å system occurring alone is recorded from the cranefly *Tipula* and a praying mantis while the occurrence of both 31 Å and 41 Å repeats in the same animal are documented for the moth *Attacus atlas,* the locusts *Schistocerca* and *Anacridium* and the cockroach *Blaberus.*

Problems in Chitin Microfibrillar Arrangements

One of the most intriguing, yet not entirely resolved, problems in integumental biology is that of chitin fibre architecture, the regular

(?) geometry of fibres in three-dimensional space. This is of great importance because the procuticle is a secretory product of the epithelium and questions arise as to whether the resulting structures represent self-assembly crystallization phenomena, or whether the observed geometry is in fact both produced and controlled by the synthesizing epidermal cells.

The more immediate problem, which has been extensively studied — particularly in crustaceans since the time of Schmidt, W., has been to resolve the structural arrangement of chitin microfibrils as they occur in the intact cuticles of insects and other arthropods. The attendant arguments and evidence advanced thus far have proved as contentious and stimulating for those interested in cuticle as has the honeybee dance language controversy for students of apiculture and animal behaviour.

Following the introduction of electron microscopy and the subsequent publication of literally hundreds of both transmission and scanning electron photomicrographs of cuticle, we must still consider the well-known photomicrograph illustrating the very familar parabolic patterning of chitin microfibrils first described from insect cuticle by Locke, M.. How ought these patterns to be understood? In a major extension of the ideas of Schmidt, W., Bouligand, Y. proposed a geometrical interpretation to explain the parabolic patterns and commonly encountered in arthropod cuticles.

The Bouligand model basically asserts that chitin microfibrils occur in sheets within the cuticle. Within any given sheet all of the microfibrils are parallel to one another and, often, are also parallel to the surface plane of the piece of cuticle within which they occur. (However, there are numerous exceptions to this latter condition where microfibrils curve in arcs with reference to the surface plane of cuticle, and this fact provided the basis for the early model of Locke, M..) Neighbouring sheets of microfibrils are rotated with respect to one another and always in the same direction of rotation. The *angle* of rotation may be variable or fixed in different cases. Any collection of sheets that has rotated through 180° constitutes a lamella. The *apparent* thickness of a lamella is susceptible to change depending upon the degree of obliquity with which the piece of cuticle was cut on the ultra-microtome. In this view, parabolic patterns of microfibrils arise as optical artefacts, the genesis of which is apparent from a consideration of the Moire pattern.

We have recapitulated the *ideal* case in the above model.

However, there are a number of exceptions and anomalies which have not yet been satisfactorily explained in terms of Bouligand's helicoidal model. The most important of these is the occurrence of what seem to be unequivocal vertically disposed microfibrils, not to be confused with pore canal filaments. To overcome this difficulty WeisFogh, T. proposed a different model in which the microfibrils are continuously secreted and the sheets of microfibrils gradually change position vertically. In this way, a microfibril present in one region of a helicoid turns slightly and can therefore contribute to more than one apparent sheet of microfibrils. This is usually termed the "screw-carpet" model, and is analogous to crystal structure in some synthetic plastics. The general applicability of this model has never really been explored to any extent. It is certainly worthy of further consideration.

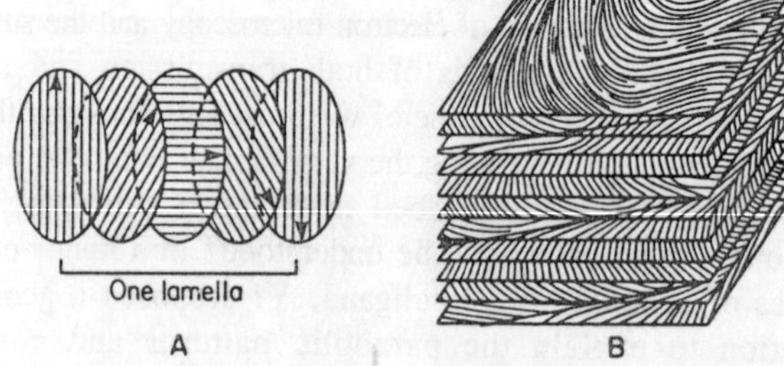

Fig. 2.8. A: Diagrammatic representation of only a few sheets of chitin microfibrils rotating through 180° and defining a single lamella. The arrows indicate the orientation of the microfibrils within each of the sheets. B: A variation of A in which the chitin microfibrils of any one sheet are curved instead of being parallel.

Fig. 2.9. Moire pattern used by Bouligand, Y. to demonstrate how the parabolic pattern of arcs arise in sections of cuticle viewed electron microscopically.

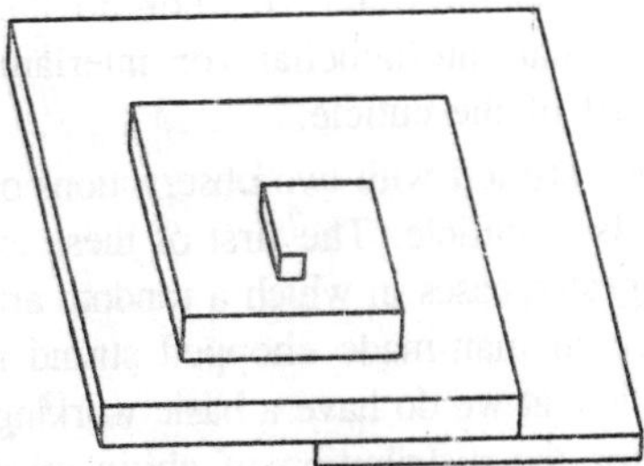

Fig. 2.10. Diagrammatic representation of "screw-carpet".

There has been considerable doubt expressed as to the reality of helicoidal arrays in many cases. A number of arguments have been advanced against the Bouligand model, and that of Weis-Fogh by implication, as follows. Both helicoidal models preclude the possibility of separate lamellae or laminae of microfibrils, yet evidence for the existence of discrete layers has been provided from numerous decapod crustaceans. Geometrical considerations of both the Bouligand and Weis-Fogh models require that the artefactual laminae of microfibrils actually spiral upwards, so that what are only *apparent* laminae would be out of register somewhere around four faces of a truncated pyramid. Dalingwater, J. has published a photograph of *Cancer* cuticle in which continuity of laminae around the angles of the pyramid is clearly discernible. In the helicoidal hypothesis there must be a discontinuity at one of the four pyramidal corners; in Dalingwater's photograph there is not a discontinuity.

The helicoidal model of Bouligand fails to satisfactorily explain the occurrence of vertical microfibrils in cuticle yet these have been variously reported from both crustaceans and from insects. Finally, the helicoidal models have not been able to successfully accommodate spiralling bow-curves that occur in cuticular bumps and corneal lens cuticle of various insects. This latter microfibrillar pattern forms the basis of the mathematical arguments of Gordon, H. and Winfree, A. that the Bouligand model is not universally applicable to insect cuticles.

Most recently, Filshie, B. and Smith, D. have contributed to the resolution of this fine structural puzzle in arthropod cuticle with their studies of a crayfish, *Panulirus*. Essentially, they suggest that the reality of vertically arranged structures in the cuticle which *appear* to be microfibrils can be explained by including regular rows of particles (possibly proteinaceous) oriented perpendicularly to the chitin microfibrils. But Filshie, B. and Smith, D. state that their vertical components are certainly not the same as those claimed by Dennell,

R.; Mutvei, H.; and Dalingwater, J.; nor do they agree with the latter three authors that interlamellar (or interlaminar) microfibrils form an integral part of the cuticle.

In any event, we are left with two observations on the arrangement of chitin microfibrils in cuticle. The first of these is that there are no known or even suggested cases in which a random array of microfibrils (such as glass fibre in man-made chopped strand mat) occurs. The second observation is that we do have a basic working model, whatever its shortcomings, for the distribution of chitin microfibrils in insect procuticle as proposed by Neville, A. and Luke, B.. Bearing in mind reservations discussed above, and that chitin microfibrillar architecture is not the same thing as whole cuticular structure, the model of Neville and Luke is certainly of heuristic value.

The "two-system model" of cuticle architecture advanced by Neville, A. and Luke, B. avers that:

(1) sheets of chitin microfibrils are helicoidally arranged in the Bouligand manner in the protein matrix and/or

(2) neighbouring sheets of microfibrils may be oriented in the same direction and so form a "preferred" layer. (Preferred is not used here in the usual physical way to mean statistically dominant, but in an absolute sense.)

Neville, A. and Luke, B. claim that most exocuticles are helicoidal throughout their thickness (exhibit artefactual parabolic lamellae) with the few exceptions of some locusts, mantids and beetles, and that various permutations occur in the endocuticle (neither exocuticle or endocuticle are clearly defined by these authors). The following patterns have been described from the endocuticle of insects and are summarized by Neville, A.:

Entirely helicoidal microfibrillar sheets have been reported for Lepidoptera, Diptera, Coleoptera, Collembola, Diplura and Thysanura.

(2) Entirely preferred or unidirectional sheets occur in the femoral apodeme of a grasshopper, *Eutropidacris*.

(3) Alternating helicoidal and preferred layers in which all of the preferred layers lie in the same orientation, occur in Orthoptera, Cursoria and Mantodea; the alternation of layers is controlled by circadian clocks.

(4) Alternating helicoidal and preferred layers in which the preferred layers change direction with respect to one another (analogous to orthogonal plywoods) are found in Hemiptera, Odonata, some Coleoptera, Phasmida and some Mecoptera.

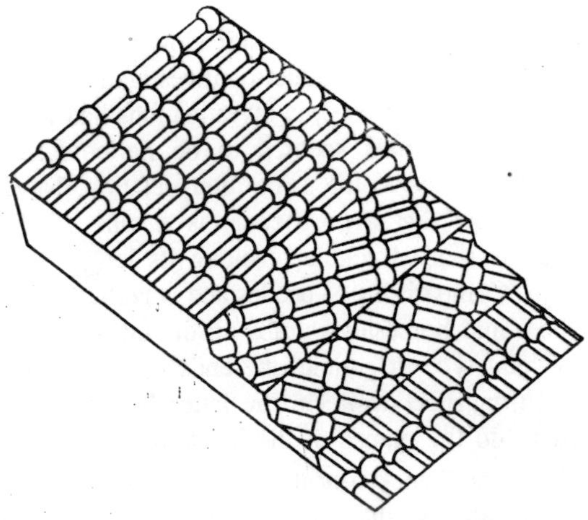

Fig. 2.11. The occurrence of apparent vertical microfibrils in the inner endocuticle of the crayfish, Panulirus.

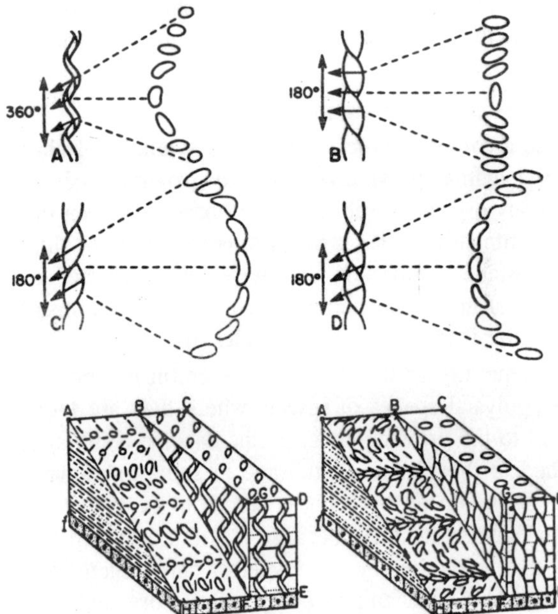

Fig. 2.12. Pore canal model of Neville, developed to relate pore canal structure to helicoidally arranged chitin microfibrils. In the surface plane (top right of diagram) the shapes and directions of pore canals are the same but change with the changing direction of the helicoid microfibril sheets.

Interestingly, this proposed classification of Neville, A. and Luke, B. can be regarded as a byproduct of investigations on the structures of pore canals. Light microscopic investigations had led early workers to infer that pore canals seemed to be helicoidal, and electron microscopic studies led to the same interpretation. Neville, A. *et al.* were able to show that a pore canal is a tube of elliptical cross-section which is twisted regularly along its length so as to form a twisted ribbon. Comparisons of beetles and waterbugs, on the one hand, and honeybees and horseshoes crabs on the other, showed that in the former, plywood type of cuticle, the ends of the ellipses were in the axis of the chitin sheet while in the latter helicoidal cases the pore canals twist as do the planes of the helicoids.

In applying the "chicken-or-the-egg" question to pore canals versus helicoidal fibre sheets, there are several examples of cuticle from which pore canals are absent but the chitin microfibrils are helicoidally arranged. While this kind of evidence is admittedly circumstantial, it seems reasonable to infer that pore canal shape results from the forces associated with crystallization of the forming cuticle and not vice-versa.

Biosyntheses and Degradaton of Chitin

While a complete description of the synthesis of chitin monomers and their eventual polymerization into growing chains has not been made for any organism, substantial advances have been made in parallel studies on fungal and insect material (Kramer *et al.,* this volume). In the first instance, sufficient data have been collected using electron microscopical and radiochemical techniques to establish that chitin is in fact synthesized by the insect epidermis.

It is generally assumed that active chitin monomers are secreted into the apolysial spaces of insects where they are then polymerized and added to the growing chain by the enzyme chitin synthetase which is attached to the plasma membrane in some unknown manner. The eventual resolution of this point is crucial to an understanding of chitin microfibrillar architecture, because it is this last step in the synthetic pathway that holds the key as to whether chitin microfibrils are formed by self-assembly mechanisms or by cellular organelles.

Some light is shed on this problem in recent studies of fungal preparations by Bartnicki-Garcia, *S. et al.*. In fungi, chitin synthetase is very probably produced as an inactive zymogen in the endoplasmic reticulum of a synthetically active epithelium. The proenzyme is contained in cytoplasmic vesicles, termed chitosomes, in which it is transported to the plasma membrane. The zymogen is then activated by proteases on the cell surface which release the chitin molecules and so eventually polymerize the chitin monomers. This account is somewhat hypothetical because there is no unequivocal proof that chitosomes actually occur in living cells. Chitosomes could be artefacts of the ultracentrifuge.

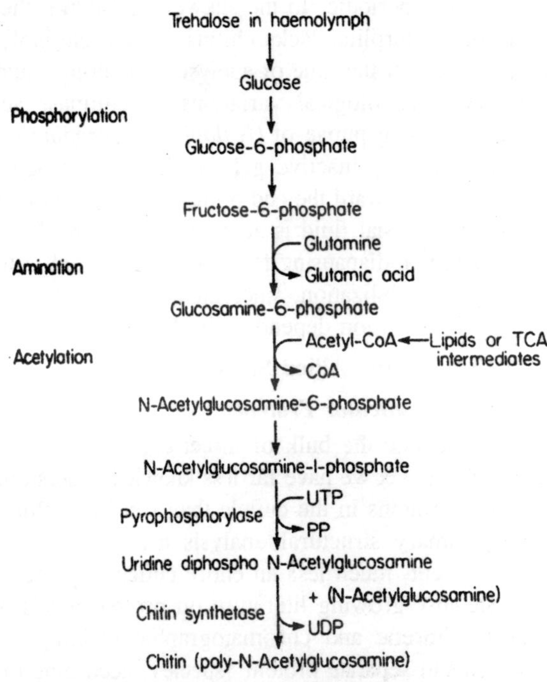

Fig. 2.13. Pathway for the biosynthesis of chitin.

As far as insects are concerned, it remains to be shown that chitin is indeed secreted into the apolysial space in monomeric form. Nacetylglucosamine residues may be presumed to be transferred from UDPAG to the forming chain. If chitin synthetase is located in the plasma membrane, the UDPAG could well come from inside the cell, with the chains being polymerized in the extracellular apolysial space; but confirmation is required. As we have said, the chemical synthesis of chitin appears to be the same in fungi and in insects. However, the fine physiological controls of biosynthesis will certainly differ since we know that chitin synthesis in insects is controlled by 20hydroxyecdysone and is also pH- and temperature-dependent.

The degradation of chitin is a normal catabolic process associated with growth and moulting in insects. Because the exuvium generally consists of epicuticle and exocuticle, most of the chitin which is digested and resorbed during moulting must be endocuticular in origin. It is digested by the enzymes chitinase and chitobiase but recent work suggests that we are dealing with a collection of as yet unpurified enzymes.

Just as the synthesis of chitin is a cyclic process, the elaboration of chitinases is also periodic. In the silkworm, *Bombyx,* the epidermis of the intermoult caterpillar lacks chitinases, and chitinolytic activity can be detected only at the time of apolysial (moulting) fluid secretion. There are other physiological variations on chitinase activity. For example, in diapausing pupae of *Hyalophora cecropia* the epidermis secretes an enzymically inactive gel which becomes activated when the gel becomes fluid toward the end of diapause while in non-diapausing Lepidoptera the apolysial fluid is active on secretion. The differences in diapausing and non-diapausing forms may be related in turn to basic differences in salt mobilization, since the activation of the chitmolytic proenzymes is calcium ion dependent in the latter.

Proteins of Cuticles and Allied Structures

Heterogeneity of Cyuticular Proteins

Proteins constitute the bulk of insect cuticle on the basis of dry mass measurements, yet we have far less knowledge about the structure and varieties of proteins in the cuticle than we do of chitin. To date, not a single primary structural analysis has been reported for an oligomeric fragment, much less an entire cuticular protein. There is, however, a steadily growing literature on protein bands obtained by various electrophoretic and chromatographic techniques and using different solvents to separate protein "species" according to their bond characteristics. For example, proteins can be classified as those which are water-soluble (these used to be termed "arthropodin" but the term means very little today) and others which are held in the cuticle by van der Waal's forces, hydrogen bonding or covalent bonding. The sequential extraction and electrophoretic characterization of proteins in this way has come to be known as the "Hackman series" after Hackman, R..

The sequential extraction of cuticular proteins is of direct importance to understanding protein structure within the cuticle, because the products of the extraction sequence indicate to what extent a given fraction might interact with other components such as chitin or proteins in the intact cuticle. But caution must be introduced in the interpretation of proteins studied in this way. Different solvents will affect different bonds in different ways, since not all hydrogen bonds, for example, are of the same strength; nor are given bonds necessarily equally accessible to a solvent. Because of the extreme sensitivity of proteins to variations in analytical technique, it is very likely that many of the bands obtained are but fragments of whole proteins so that a protein banding profile might well contain numerous oligomers.

TABLE 2.2: AMINO ACID COMPOSITION OF INSECT CUTICLES AND CUTICULAR PROTEINS GROUPED ACCORDING TO THE SECOND LETTER OF THE GENETIC CODE

Species	Second letter of the genetic code			
	U	C	A	G
	Phe,Leu,Ile, Met,Val	Ser,Pro, Thr,Ala	Tyr,His,Glx, Asx,Lys	Cys,Trp, Gly,Arg
Agrianome spinicollis				
larval cuticle	226	358	285	131
water-soluble protein				
Anoplognathus montanus				
larval cuticle	231	283	307	179
Anabrus simplex				
abdominal intersegmental connectives	218	282	325	175
Bombyx mori				
larval cuticle	185	362	310	142
Lucilia cuprina				
larval cuticle	196	303	338	162
Muscafergusoni				
larval cuticle	178	318	328	176
Pison spinolae				
larval cuticle	248	294	335	123
Xylophasia monoglypha				
abdominal cuticle	179	338	331	152
resilin				
Rhodnius prolixus				
abdominal cuticular	92	298	184	426
protein	219	446	156	179
Periplaneta americana				
adult cuticle	199	435..	236	129

Protein bands vary enormously, and heterogeneity of bands occurs within the same animal over the course of its life cycle, as in the waxmoth *Galleria mellonella* or not, as in *Schistocerca gregaria.* Protein variations certainly occur in different kinds of cuticle taken from the same animal, e.g. solid and intersegmental cuticle in *Hyalophora cecropia*; but, interestingly enough, there are no apparent differences in the cuticular protein bands of physogastrous and nonphysogastrous termite queens.

The significance of protein heterogeneity to structure and function remains an open question. Nonetheless, an origin for such heterogeneity has been proposed by Hackman, R. taking a leaf from genetic variations in insect population studies. He suggests that natural insect populations (not to mention that "laboratory artefact" *Drosophila)* are polymorphic at thousands of structural gene loci and that every individual in a large out-breeding population is likely to have a unique protein profile. The argument naturally follows that any character, genotypic or phenotypic, need only be non-lethal or neutral in terms of natural selection forces to be retained in a population. Indeed Jukes, T. has recently provided a compelling argument that evolution can be viewed as the retention of silent nucleotides with respect to structural protein composition.

Hackman, R. has noted that there is a simple relationship in amino acid coding based on the second letter of RNA, and that there is a high probability of interchange between the first and second letters of the code. Hackman, R. has compared the amino acid residues of tanned and untanned cuticles and grouped them according to the second letter of their genetic code origins. In this way it can be seen that what at first appears to be a series of entirely unique cuticles can actually be resolved into very similar profiles indeed. Thus, untanned cuticles constitute a group similar in amino acid composition and one that is markedly different from both tanned cuticles and from resilin. We can suggest that it may be the *kinds* of proteins rather than particular proteins which are important to structure, a generalization consistent with the effects of residue composition on molecular behaviour.

Before discussing the properties of amino acids and particular kinds of proteins we should mention the possible origins of cuticular proteins. There is still uncertainty as to whether proteins occurring in the cuticle are synthesized only by the epidermis and/or whether they might be produced somewhere else in the body and transported to the

Table 2.3 : Amino acid composition of various insect cuticles

Species	Residues withsmall side-chains	Highly polar residues	Proline	Alanine	Cuticle type
Coleoptera					
Agrianome spinicollis	32	43	10	—	arthrodial/caterpillar
Tenebrio molitorl	40	36	9	?	solid
Tenebrio molitor	42	36	7	?	solid
Tenebrio molitor	42	27	9	?	solid
Pachynoda epphipiata	37	45	8		arthrodial/caterpillar
Pachynoda epphipiata	38	42	10	trace	solid
Pachynoda epphipiata	50	29	7	trace	solid
Xylotrupes gideon	47	25	8	1	solid
Dictyoptera					
Periplaneta americana	43	17	10	25	solid
Diptera					
Lucilia cuprina	29	43	9	—	arthrodial/caterpillar
Lucilia cuprina	28	46	9	9	arthrodial/caterpillar
Calliphora augur	28	47	9	—	arthrodial/caterpillar
Drosophila melanogaster	26	47	5	1	arthrodial/caterpillar
Musca domestica	29	45	6	8	arthrodial/caterpillar

Species	Residues withsmall side-chains	Highly polar residues	Proline	Alanine	Cuticle type
Hemiptera					
Rhodnius prolixus	53	9	8	—	arthrodial/caterpillar
Lepidoptera					
Bombyx morn	31	47	10		arthrodial/caterpillar
Bombyx mori	21	38	10	14	arthrodial/caterpillar
Sphinx ligustri	28	51	9	'	arthrodial/caterpillar
Sphinx ligustri	38	41	12	9	solid
Sphinx ligustri	44	39	8	?	solid
Hyalophora cecropial	29	50	10	?	arthrodial/caterpillar
Aglais urticae	29	52	9	9	arthrodial/caterpillar
Xylophasia monoglypha	40	38	7	+	solid
Odonata					
Aeshna juncea	64	20	7	—	resilin
Orthoptera					
Schistocerca gregaria	63	20	7	—	transitional

cuticle through the haemolymph. While it is often tacitly assumed that cuticular proteins are synthesized by the epidermis there is close similarity between some of the proteins in the haemolymph and cuticle in the moth *Manduca sexta* and in *Locusta migratoria*, which implies the possibility of transport. Similarly, radiolabelled proteins injected into label-free insects can subsequently be recovered from the cuticle.

Amino Acids

Differences in the amino acid compositions of different kinds of cuticles from different insects of varying age are well established in the literature and have been discussed in considerable detail . Rather than retabulate this vast amount of data, we rather enquire as to the relationships between amino acid composition and properties versus structure and function in intact cuticles. A few generalizations can be made in this regard.

The protein fractions of untanned cuticles from holometabolous larvae are usually rich in generally hydrophobic residues with bulky side chains. Such residues are associated with loose molecular packing in structural protein chemistry. Thus, the resulting packing is less crystalline and this property imparts greater flexibility to larger structures where such residues occur. There are also several suggestive mechanical correlates: cuticles that are untanned but rich in bulky side-chain residues are often tough materials, but are of relatively low strength and stiffness. Similarly, loose packing in a cuticle coincides with increased creep, and this property could indeed facilitate intrastadial growth in caterpillars. A corollary to this is that cuticular protein hydrolysates in which there is a preponderance of residues with small side-chains implies closer chain packing and a corresponding increase in stiffness and strength.

On average, the smaller amino acid residues predominate in the hydrolysates of sclerotized cuticle. There is also a correlation between the relative abundance of non-polar residues and relative cuticular hardness. Similarly the number of free amino groups available for substitution decreases as sclerotization progresses.

Returning to the scheme of Hackman, as outlined in Table 2, one observes the relationships that group A contains the hydrophilic amino acid residues and that sclerotized cuticles are consistently low in such residues.

Andersen, S. recently proposed some very interesting extensions of these ideas, and points out that "hydrophobic and hydrophilic" are boundary terms defining widely ranging behaviour of residues with

Table 2.4 : Hydrophobicity indices of various insect cuticles according to their amino acid composition

Species	HI
Coleoptera	
Pachynoda epphipiata larva	4
pupa	32
imago	16
Tenebrio molitor larva	30
pupa	32
imago	34
Diptera	
Lucilia cuprina	
larva	– 4
puparium	– 13
Calliphora vicina	
larva	– 4
Calliphora augur	
puparium	– 24
Lepidoptera	
Bombyx mori	
larva	9
Sphinx ligustri	
larva	– 19
pupa	27
imago	27
Dictyoptera	
Periplaneta americana	
imago	41
Isoptera	
Macrotermes bellicosus	
physogastric queen, pleural membrane	– 38
neotergites	1
Hemiptera	
Rhodnius prolixus	
fifth-instar larva, abdomen	59
Orthoptera	
Locusta migratoria	
imago, pronotum	55

Species	HI
Schistocerca gregaria	
fifth-instar larva, femur	51
imago, femur	52
tibia	51
ventral thorax	53
abdominal terga	36
abdominal intersegmental membrane	– 5
membrane of coxal cavity	– 10
wing hinge	– 12

respect to hydrophobicity. Taking calculated values for the energy changes associated with the transfer of a given residue from water to ethanol, a hydrophobicity index of amino acids can be obtained. Hydrophobicity indices for whole cuticles are obtained by extrapolation from the relative percentage contributions of different residues. While some simplifying assumptions have been made (for example, the acid amide residues may be either amides or free acids) a meaningful indication of the relative strengths of hydrophobic interactions between protein molecules is still obtained.

It can be seen from Table 4 that there is a spectrum of values from + 60 to – 40, and that there is no sharp division of cuticle types along the way. It is of interest to note that stiffness and strength of cuticles also cover a spectrum and that very stiff cuticles have a high positive index, those of low strength and stiffness a more negative index. Citing the example given by Andersen, S., unsclerotized cuticles from *Bombyx mori* and *Locusta migratoria* both appear to lack covalent bonding in the protein matrix so that when samples of these cuticles are stretched, resistance to displacement will derive principally from the chitin microfibrils. This implies that the elastic modulus (stiffness) will indicate an approximate measure of the strength of the reactions. One can conclude that a general consistency appears between the mechanical properties spectrum on the one hand, and both hydrophobicity index as well as the grouping of amino acid residues according to the second letter of the genetic code.

The Proteins of Eggs

Insect oothecae consist primarily of proteins. They occur widely amongst the Cursoria, Mantodea and Orthoptera and are also secreted to a lesser extent by Coleoptera and Lepidoptera. The ootheca of the cockroach, *Blatta orientalis* is crystallographically related to the extended f-fibroins and is central to the history of tanning in insects in that they were used in what is considered to be the first modern chemical analysis of tanning by Pryor, M.. Oothecae are usually secreted from paired collateral glands of the abdomen as proteins in solution which then crystallize as the secreted materials come together outside of the animal.

The proteins of the praying mantis ootheca are also secreted in liquid form, but the final product consists of a series of highly oriented ribbons. The ootheca of the tortoise beetle, *Aspidomorpha,* has recently been studied both by diffraction and electron microscopic techniques and found to consist of highly ordered tactoid bodies which contain a microfibrillar structure. The conformational affinities of *Aspidomorpha* oothecal protein are uncertain; however it has mechanical similarities to feather keratin. Little more can be said about oothecal protein structures other than that they are rich in the polar residues glycine and tyrosine, and rather low in cysteine and methionine. These last observations, *vis-a-vis* solid cuticle, would make comparative hardness indices and hydrophobicity values for oothecal material extremely interesting, to say the least.

The insect egg shell is a multiple-laminated structure synthesized by the epithelial cells of the ovarian follicles and is, like the cuticles of insects, a physical record of the activities of the cells that produce it. The egg consists of a chorion, vitelline membrane and oolemma. The chorion has generally been considered to be the principal structural layer of the egg, conferring toughness and rigidity to the whole. More recently it has been suggested that the vitelline membrane also confers structural stability to eggshells, on the grounds that dechorionated eggs maintain their shape.

The principal structural protein of the insect egg is termed chorionin and this material has been analysed chemically, microscopically and crystallographically for many orders of insects. The chorionins of numerous species have been fractionated and, following the tradition in keratin chemistry, have been classified as type I or type II chorionin proteins depending upon either a high or a low sulphur content. Similarly, the amino acid compositions of various species of eggshells have been reported. The same difficulties apply to the interpretation of egg proteins as to cuticular proteins: degradation by chemical analytical techniques probably produced collections of oligomers so that we cannot state unequivocally whether "chorionin" is a protein or a collection of proteins.

TABLE 2.5 : AMINO ACID COMPOSITION OF PROTEIN FRACTIONS FROM INSECT EGGSHELLS

Species	*Bombyx mori*			*Antheraea pernyi*		*Sympetrumin fuscatum*			*Tenebrio molitor*			*Rhodnius prolixus*		*Oryctes rhinoceros*	*Gryllus mitratusb*	
Fraction	I	II	III	I	III	I	II	III	I	II	III	I	II		I	III
Cys	1	3	14	5	13	1	3	58	46	44	32	16	25	41	6	19
His	4	0	6	0	4	0	0	71	9	12	10	38	32	18	12	22
Arg	23	28	14	22	18	6	6	3	38	23	10	27	32	20	18	27
Asp	39	42	53	36	44	41	22	17	81	73	16	31	53	69	49	36
Thr	32	17	28	28	27	33	23	6	41	27	22	54	41	32	10	21
Ser	31	39	37	38	47	80	56	11	48	71	113	12	32	69	237[7]	50
Glu	36	30	64	43	45	100	11	30	60	81	33	33	70	70	105	35
Pro	42	20	56	36	30	60	72	120	62	87	169	49	59	67	13	107
Gly	361	302	101	324	137	152	340	245	68	61	48	226	142	79	21	66
Ala	83	27	46	127	60	73	15	158	63	50	137	80	34	61	16	70
$\frac{1}{2}$Cys[6]	57	256	5	59	7	180	218	0	29	36	0	49	147	35	12	2
Val	60	44	33	62	38	21	8	9	65	55	29	73	82	68	77	31
Met	0	3	7	3	8	1	6	17	8	9	10	0	1	3	trace	6
Ile	32	18	21	38	24	14	13	5	48	43	12	40	11	39	37	14
Leu	64	12	39	69	45	46	39	7	75	69	15	12	26	50	22	30
Tyr	79	13	12	73	25	31	34	103	35	21	27	84	20	43	6	21
Phe	19	10	14	14	14	0	0	3	31	26	4	22	14	19	6	20
Try	18	7	4	7	9	5	9	trace	2	1	0				17	12
Dityrosine						0	0	4	0	0	2.1				0	0.6
Trityrosine						0	0	1	0	0	0.7				0	0.2
Amide	32	30	76	29	62	121	19	16	68	78	40	61	118	70	34	35

The eggshell consists of an intrachorionic space near the inner margin of the chorion, and is accessible to the external atmosphere via aeropyles which traverse the chorion. It has been suggested by Hinton, H. that the position of the air space shortens the path of oxygen to the developing embryo, and that water loss is reduced through the limited cross-sectional area of the aeropyle. However, physiological measurements in support of this conjecture are wanting. There is considerable variation in the surface properties of eggs and the distribution of aeropyles in insect eggs. This is somewhat surprising in view of the chemical similarity of egg chorionins.

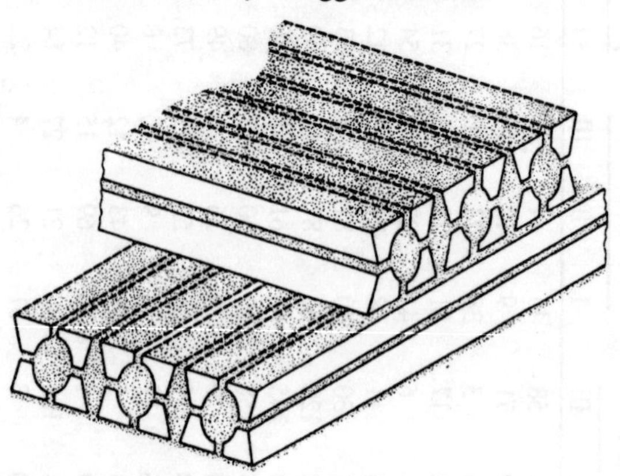

Fig. 2.14. Cross-ply arrangement of fibres and impregnating matrix (stipling) between fibres as reconstructed for the chorion of the beetle, Oryctes.

Recently electron microscopic studies summarizing data from many different insects as to the structure of the chorion were published by Furneaux, P. and MacKay, A.. For beetle chorions that reflect a highly ordered type of structure these authors have suggested a crossply structure of fibres embedded in a more amorphous matrix substance, presumably also of protein. They envisage that each of the fibre bundles consists of four microfibrils of about 40 A diameter. For other eggs such as those of the moth *Hyalophora cecropia,* Smith, D. *et al.* have reported that the microfibrils are arranged in helicoidal lamellae.

The vitelline membrane appears to be both more variable, and less understood, than is the chorion. It ranges from negligible as a percentage of the total egg in the grasshopper, *Melanoplus differentialis,* to nearly four-fifths of eggshell mass in the dragonfly, *Sympetrum infuscatum*. The vitelline membrane is the site of the principal wax layer in eggshells and the wax is formed after the bulk of the shell has been synthesized. A putative arrangement for the vitelline membrane from serial sections has been prepared by Furneaux, P. and MacKay, A.. Despite the recent appearance of H. Hinton's large work on many aspects of insect eggs, the subject must be regarded as in its infancy.

Resilin

The discovery of resilin by Weis-Fogh, T., and its subsequent characterization by that author and his colleagues, represents what must certainly be regarded as an intellectual *tour-de-force* in the first rank of insect structural studies. It was initially characterized from the prealar arm and wing-hinge ligaments of the locust, *Schistocerca gregaria,* and from the pleuro-subalar muscle tendon of the dragonfly, *Aeshna grandis*. The recoverable mechanical deformability of these preparations, in contrast to sclerites, led to the distinction between "rubber-like" cuticle and "solid' cuticle.

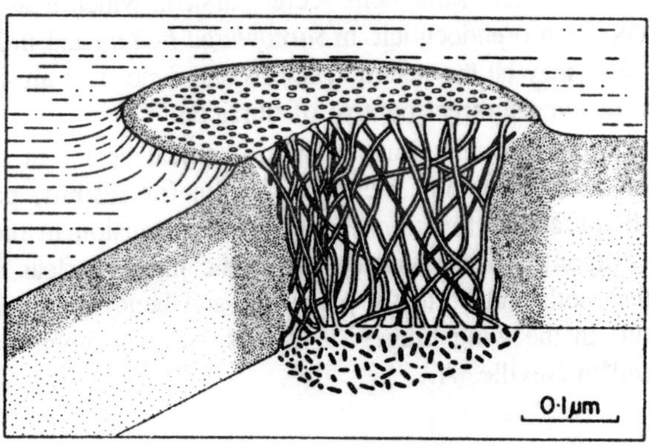

Fig. 2.15. Twisted and intertwining tubules open to the surface. These tubules are surrounded by an electron-dense (stippling) "sleeve". It is suggested that this system is responsible for wax transport in the egg shell.

Pure resilin is both colourless and transparent, and reveals no evidence of structure in electron micrographic preparations. It is optically and mechanically isotropic but does exhibit strain birefringence in the direction of deformation. When solvated it can be distended reversibly over 200% of its initial length, held to the stretched length by drying out, and when resolvated and released will return to its original rest position. Its principal physical properties are its long-range elasticity, elastic efficiency (which is an index of recoverable energy obtained when deformation is discontinued), and lack of either creep or stress relaxation.

Weis-Fogh, T. predicted that resilin ought to show molecular structural similarities to botanical rubbers in which there is a threedimensional network of chains nearly free of or another, thermally agitated, randomly kinked an fixed within a network by only a very few stab cross-links. Andersen, S. subsequently characterized these necessary covalet cross-links as new fluorescent amino acid dityrosine and tertyrosine, which ultimately deriN from tyrosine. Amino acid analyses of seven preparations from different species indicate quite close overall similarity in composition.

On the basis of composition, resilin differs mai kedly from other structural proteins in that it is ric in glycine, hydrophilic residues, and proline bt. completely lacks hydroxyproline, tryptophan or th sulphur-bearing residues. Resilin remains unsequenced but a very likely car didate for future sequencing studies of inset proteins. Though its biosynthesis also remains to b established, the deposition of resilin is certainly lin ked to the median neurosecretory cells of the brair and it is also known that these same cells secret bursicon which is associated with the deposition o endocuticle in *Sarcophaga bullata* and in *Locust, migratoria* so it may be tha resilin synthesis and deposition are likewise unde direct or even indirect control of bursicon.

In insects, resilin functions as a mechanical sprint in tension, compression and in bending members Resilin (or resilins) probably occur in all insects. It functions as a spring in tension in the flight system of dragonflies; in compression in the jumping fleas and the recoil of the moth and butterfly proboscis and bending (tension and compression) in the wing hinges of locusts. Several other examples can be found in Neville, A..

Silks

(a) *Structure* Protein silks are elaborated by virtually every order of insects in which they have been sought. In addition to the familiar cocoon formed by spun fibre of circular cross-section, protein silks also occur in ribbons and sheets and are used in myriad ways: for anchoring dragonfly eggs to rocks in fast streams or as egg stalks in lacewing flies, as food-traps for caddis-flies and tunnels for webspinners, and in very many other ways. Aside from these special uses in some life stage of an insect, silk holds importance for the study of the integument because it is produced by an epithelium (if not always the epidermis) and because it is vastly more accessible to crystallographic and chemical study than are proteins occurring *in situ* in a cuticle. Silk affords us much insight into the behaviour of structural proteins, and as to how behaviour can be related to molecular structure.

There is another point of interest in silks. It will be seen that

Table 2.6 : Amino Acid Composition of Various Resilins.

	Schistocerca gregaria					*Aeshna juncea* flight tendon	*Oryctes rhinoceros* abdominal spring
	Wing hinges		Prealar arms		Clypeo-labral spring		
	a	b	a	b	b	b	a
Aspartic acid	9.9	10.2	10.8	11.3	7.5	9.4	22.6
Threonine	3.1	2.9	3.0	3.1	3.4	2.0	4.1
Serine	7.9	7.7	8.1	7.5	8.9	12.8	4.9
Glutamic acid	4.6	4.3	4.8	3.8	6.4	4.2	5.9
Proline	7.6	7.5	7.9	7.7	7.2	7.5	8.9
Glycine	39.1	41.1	36.7	39.7	31.4	42.2	33.5
Alanine	11.2	11.2	10.6	11.2	13.5	7.0	4.6
Valine	2.6	2.3	3.2	2.5	3.1	1.2	2.5
Cystine	—	—	—	—	—	—	—
Methionine	—	—	—	—	—	—	—
Isoleucine	1.7	1.3	1.7	1.4	1.9	0.9	2.6
Leucine	2.3	2.2	2.4	2.8	5.2	3.0	3.9
Tyrosine	2.5	2.1	3.1	1.2	1.9	1.2	0.5
Phenylalanine	2.6	2.5	2.4	2.4	2.8	2.1	2.2
Lysine	0.5	0.4	+	0.5	0.9	0.8	0.5
Histidine	0.9	0.8	5.3	1.1	0.9	1.2	0.5
Arginine	3.5	3.6	—	3.8	5.1	4.6	2.9
Sum of polar residues	32.9	32.0	35.1	32.3	35.0	36.2	41.9

silks are relatively distensible in fibroin form. However it has very often been suggested that one of the main "reasons" why an insect must moult is owing to the inextensible nature of a-chitin. It should be pointed out that all of the structural proteins are relatively distensible compared with chitin, but this is not so when they are tanned in a cuticle. It may well be that tanning is required to impart stability to the structural protein, and this could be a "reason" why insects moult.

The discovery of transconformational changes in keratin established that a protein polymer could exist in at least two different configurations, a folded a and an unfolded parallel β form. This landmark in molecular biology was given important extension by Rudall, K. who discovered that epidermal protein from the cow's lip could be thermally supercontracted into yet another stable configuration, the cross-β form. Rudall also established experimentally the relationships for the KMEF (keratin, myosin, epidermin and fibrinogen) group of structural proteins.

All three of these protein configurations were subsequently discovered and characterized from naturally occurring insect material. Thus, parallel-β fibroin in Tussah silk from the caterpillar *Antheraea pernyi* was described by Marsh, R. *et al.*; the α-helical configuration was found in the honeybee cocoon silk and the cross-βpattern in the egg stalk of the green lacewing flies *(Chrysopa)* by Parker, K. and Rudall, K..

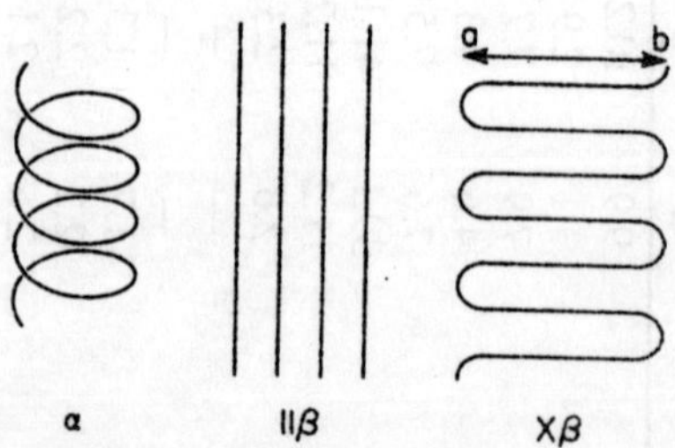

Fig. 2.16. Highly diagrammatic representations of fibroin crystal structures.

The β-structure consists of highly oriented fibres in which the crystallites are arranged with their long axes in the same direction as the fibre axis. The ribbon-like crystallites are linked by short bends so that a continuously folded polypeptide chain results. These chains lie perpendicular to the ribbon surface and parallel to the long axis of the ribbon. The structure of the a-helical pattern characteristic of honeybee silks, as reconfirmed by Atkins, E., is seen as a crystal structure in which four helices coil upon one another, much like threads in a piece of cotton, and parallel to the fibre axis. The β-pleated sheet configuration is the well-known form in which the crystallographic β-axis is coincident with the fibre axis.

The distinctness of structure in the different configurations as well as the behaviour of such fibroins, greatly depends upon the secondary structure of the protein. That is, the structure depends upon the hydrogen bonding topologies, and the transconformations which can occur between them require the disruption of the secondary protein structure through changes in the positions of the crystallites with respect to their position and arrangement within the fibre.

(b) *Problems of crystallinity* At the time of writing the amino acid compositions of several hundred different insect silks have been described, many of which have been tabulated in the review of Lucas, F. and Rudall, K.. In the quest to relate amino acid composition of cuticular proteins to structure of cuticles and to functional properties, relationships involving "kinds" of amino acids and their relationships to water emerge; this is analogous to the hydrophobicity index with resj to mechanical properties suggested by Anderset.

It is well known that the degree of crystallir the ratio of ordered or crystalline to disorderec amorphous regions, greatly affects the 1: material properties of a polymer, especially in to of the relative availability of chemically reac sites. Moreover, the average crystalline content decrease with increasing fibre diameter. Some estimate of the degree of crystallinit fibroins can be made from a consideration of air acid composition. TI amino acids with long side-chains tend to occu the amorphous regions, and those with short si chains in the crystalline regions when the unit ccc the *b* direction is small. Since only glycine, alan threonine and serine occur in highly crystalline fi tions of *Bombyx* fibroin these specific residues are taken as short side-ch standards for purposes of comparison.

Lucas, *F. et al.* used this approach estimate the degree of crystallinity for fibroiw *Anaphe, Antheraea* and *Bombyx,* and we h similarly calculated short side-chain-long si chain (SC/LC) ratios for *Apis, Chrysopa* and *leria* from the data of Lucas, F. and Rudall. Inspection of which immediately establishes honeybee as being anomalous because this silk has a very glycine content yet is known to be relatively crys line. By the same token, it is also possible for 1, side-chain residues to occur in crystalline regi (e.g. braconid and spider fibroins). A possibly more useful approach for estin ing relative crystallinity is suggested by studies of cellulose bast fibres.

TABLE 2.7: RELATIVE CRYSTALLINITY OF α, CROSS-β, AND PARALLEL-β FABROINS ASSESSED IN TERMS OF ELASTIC MODULI AND AMINO ACID *RESIDUES*

Specimens	(SC/LC)	(E_{Wer}/E_{ary})
Parallel-β		
Anaphe	23.4	0.81
Bombyx	6.6	0.64
Antheraea	3.9	0.32
Galleria	3.0	0.18
Cross-β		
Chrysopa	10.9	0.29
α		
Apis	~1.5[4]	0.53

The argument from cellulose states that the degree of crystallinity will be reflected in the sensitivity of the material to solution effects. For example, water can penetrate amorphous regions in a capillary manner, thereby diminishing the interactions between crystallites, or alternatively, actually compete for potential hydrogen-bonding sites within the fibre. In the work on cellulose materials it was assumed that the effect of hydration was to loosen the interaction between neighbouring crystalline regions, so reducing stiffness. It was further assumed that the modulus of the dry material approached that of crystalline cellulose, as a result of which the ratio of modulus wet to modulus dry provided an approximate index of the degree of crystallinity, a ratio of 1 indicating complete, and smaller values progressively less, crystallinity. When the six fibroins are examined for hydration sensitivity, expressed as the ratio of elastic modulus of wet to that of dry fibre, entirely different estimates of crystallinity are obtained. Indeed, a value of 0.53 for honeybee silk shows that this fibre is rather crystalline and not dissimilar to *Bombyx,* results that are consistent with crystallographic estimates for the same materials.

The importance of the insect silks in the study of fibrous proteins is that they provide relatively simple preparations for investigations of conformational changes. We have seen in the tensile stress-strain curves of the waxmoth silk, and of the other parallel-β fibroins, what can largely be described as linear behaviour in the dry state. Differences within this group are related to variations in crystallinity as estimated from ratios of wet to dry moduli, and to a lesser extent the SC/LC ratios. Honeybee silk, on the other hand, is a-helical in nature and its ability to undergo transconformational changes is well documented. Thus highly extended specimens of honeybee silk provide diffraction patterns, the main features of which are practically identical to those of *Bombyx* or *Galleria*. The same is true of highly extended *Chrysopa* silk.

If we assume from the diffractional data of Parker, K. and Rudall, K.; Rudall, K.; Atkins, E.; and Geddes, A. *et al.* that extensions of the required magnitude do result in diagrams that are indistinguishable in their main features, then it could be expected that similar extensometric relationships would occur in measured deformations of these configurations. Silks tested in distilled water closely approximate the expected behaviour. Here, all three configurations have been plotted to the same scale. However, the origins of the honeybee and waxmoth curves have been zero-shifted to the right along the abscissa in order to compensate for intrinsic differences in total extensibility. If, for example, the a fibroin were a cross-β fibroin it would require an additional 0.68 strain to achieve a total strain of 2.09 (about 700%). Similarly, the parallel-β form is zero-shifted to the right by a strain of 1.4 with respect to the cross-β form, and by 0.68 with respect to the a form.

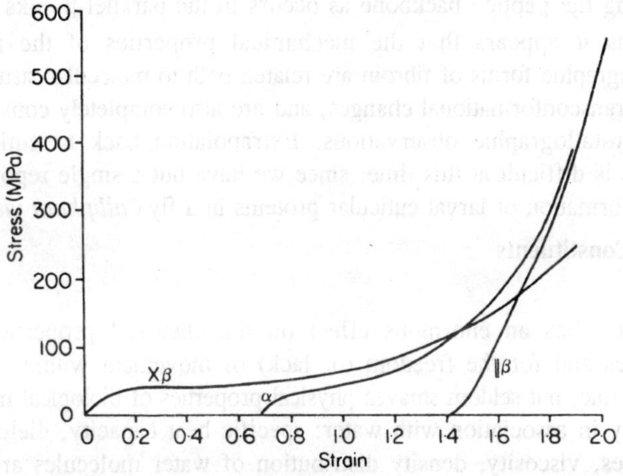

Fig. 2.17. Superposition of tensile stress-strain curves of honey-bee α-helical, waxmoth parallel-β and lacewing fly cross-β silks. Th honey bnee and lacewing fly curves exhibit low starin yield points associated with trasconformational change. Both the honeybee and the waxmoth curves have been zeor shifted to the right to illustrate the goodness of fit for the cross-β → parallel-β and α → parallel-β transconformations.

The results obtained by shifting the a-helical and parallel-β silks with respect to the cross-β lacewing silk extensibility data obtained on experiment agree within 10% of the deformations that are required to distort similarly accurately scaled molecular models. This observation enables us to make reasonable estimates of the relationship between fibre crystal structure and the measured mechanical properties of the three crystal systems of the silk fibroins. The deformation of the parallel-β fibroins is largely limited to the direct stretching of the peptide backbone, and hence stiffness reflects the deformation resistance of the primary bonding with total extensibility (including initial "slack") being further affected by the extent of bulky side-chain residues present.

The α and cross-β silks share similarity in that both exhibit a very low strain yield, a long region of transformation with finally increasing stiffness. The initial deformation behaviour of these kinds of fibroins involves the stretching and then breaking of the secondary hydrogen-bonding lattice of the crystals followed by gradual unfolding of the molecules. The unfolding phase is the low stress, plateau area which is the greater in the cross-β silk because it is more folded than the a silk. Once the unfolding has been completed the continued stretching of the fibre is essentially a process wherein the now parallel-β form has appeared, and the deformation process becomes one of stretching the peptide backbone as occurs in the parallel-β silks.

Thus it appears that the mechanical properties of the three crystallographic forms of fibroin are related both to molecular structure and to transconformational changes, and are also completely consistent with crystallographic observations. Extrapolation back to cuticular proteins is difficult at this time, since we have but a single report on the conformation of larval cuticular proteins in a fly *Calliphora vicina.*

Other Constituents

Water

Water has an enormous effect on the chemical properties of structures and for the freedom (or lack) of movement within them. Innumerable, but seldom stuavea physical properties of biological matter will vary in association with water: specific heat capacity, dielectric properties, viscosity, density distribution of water molecules around other molecules, dissociation characteristics and, of the greatest importance to the structure and performance of insect cuticles, the hydrogen bonds between liquid water and structural molecules. All of these things will also vary with temperature. The significance of water to integuments is supported in experimental measurements; however, interpretations of observed behaviour largely come from inferferences on the physics of water.

While we have no information on the molecular configurations of water in cuticles, some generalizations can nevertheless be made. Water content of cuticle varies greatly; for example, from 15% in the femora of *Locusta* to 55% in the larval cuticle of *Sarcophaga*. By extrapolation from mechanical behaviour it has been suggested that water can be present through either capillary or molecular sorption.

Physical properties such as stiffness and strength of chitins, whole cuticles and silk fibroins increase on dehydration while distensibility decreases. All three of these materials exhibit anisotropy in swelling and the direction of size increase is generally normal rather than parallel to the axes of the structural fibre. The same is true for sclerites because the fibres in the procuticle generally run parallel to the surface.

The long-range elasticity of proteins such as resilin and fibroin are solvation-dependent as is the plasticization of the abdominal cuticle of sucking bugs like *Rhodnius*. Conversely, it is well documented that increasing density and cuticular hardening go hand-in-hand with dehydration; but one must guard against thinking that dehydration is sufficient to account for hardening. This important distinction was established over three decades ago in studies of pupariation by *Sarcophaga* and most recently reiterated in studies of *Locusta migratoria* by Andersen, S..

Thus, skeletal rigidity is largely achieved through tanning or hardening processes, and partial dehydration of the procuticle is a natural consequence of stabilization (Andersen, this volume). Great skeletal flexibility, as found in caterpillars and arthrodial membranes, is associated with a high water content and perhaps relatively few covalently bound proteins in the matrix.

Pigments

Whatever the significance of pigments in insect biology may be, at the outset it is important to state that there is no correlation between the biochemical events that lead to hardening and those which lead to darkening. That very dark cuticles are often hard is merely coincidental. Although the term "melanization" has been in use for many decades to describe "darkening" the actual chemical recognition of melanin is recent indeed: melanin as such has been described as conjugated with protein in the puparium of the *fly Lucilia cuprina* or as a discrete layer of material immediately below the epicuticle in *Sarcophaga bullata*.

There are numerous external factors which apparently influence the distribution of melanins in cuticles. For example, melanin content may vary with temperature, as in *Sarcophaga,* population density of animals in a particular place, as in the locusts *Schistocerca* and *Locusta,* and in certain genetic mutants of *Drosophila.* It is likely that melanin formation is mediated by bursicon.

Black pigments also occur as granules in some insects, the larvae of the butterfly *Papilio* and in the toad bugs *Gelastocoris.* Virtually all of the myriad pigments which colour insects occur as cytoplasmic granules in the epidermis, fat body, haemolymph or other organs beneath the cuticle.

Lipids

Despite the numerous papers that have been published in the last decade or so, describing cuticular lipids of insects, a severe difficulty surrounds the interpretation of such information. This problem was recently put in perspective by Gilby, A. who noted that very little is actually known about the location of lipids within cuticles. Thus, it is tacitly assumed that the lipids discussed in section 2 are in fact epicuticular in origin.

Few studies have distinguished the cuticular compartments from which lipids have come. Among these, Locke and his colleagues have documented the transport of waxes or their precursors to the cuticular surface via pore canals and even finer, arborescing wax canals in the caterpillar, *Calpodes ethlius.* As for the procuticle, the possible significance of lipids to matrix stabilization was first mooted by Wigglesworth, V. and he has only very recently touched upon this problem experimentally.

Combining histochemical and electron microscopical techniques in the study of *Rhodnius prolixus,* Wigglesworth, V. was able to show that the lamellate endocuticle of this bug exhibits the parabolic pattern so common to insect cuticle. In addition, he has described a series of lipid laminae of the same spacing and superimposed upon the chitin-protein lamellae. These lipid laminae are quite distinct from pore canals, the walls of which also contain dense concentrations of lipid. Wigglesworth has postulated that this procuticular lipid fraction might initiate or control the deposition of the helicoidal lamellae if formation of the latter is a crystallization process as envisaged by Bouligand, Y..

These very few observations suggest many possibilities, the most important of which is that lipids might contribute to the stabilization of the procuticle by being incorporated in some of the polymerization reactions that occur there. Indeed, the suggestions, that polymerization of lipids can be viewed as analogous to the polymerization of hydrocarbons or of isoprenoid units into materials akin to nylon or polyethylene hold exciting possibilities for the future.

Inorganic Salts

The occurrence of any appreciable amounts of inorganic salts in the cuticles of insects can be regarded as relatively rare. Calcareous deposits in the form of calcium carbonate have been recorded from the external surfaces of the larvae of some Stratiomyidae and Psychodidae and from the procuticles of some Lepidoptera and Hymenoptera. Such salts as calcium phosphate have also been recorded from the puparium of *Rhagoletis cerasi* and from the puparium of *Musca autumnalis* and *Musca fergusoni* and are thought to derive from the Malpighian tubules of the larval instar which forms the pupa, a process which appears to be under control of the moulting hormone.

The eggs of several genera of phasmids have been variously reported to contain calcium salts as carbonate, citrate and oxalate. How, if at all, these salts are structurally integrated in the cuticles or eggshells of insects is not known. As to their functions we can only make inferences from Crustacea wherein calcification results in compressive strengthening of the skeleton. Perhaps this property would be desirable for fly pupae that occur in unstable substrates. Because inorganic salts are so rare in insects *vis-a-vis* Crustacea this suggests some very fundamental differences in the metabolism of these two related groups, and one worthy of further detailed investigation.

SOLID CUTICLES

General Properties

A random sample of cuticles taken from the various developmental stages of an insect's life history will show great variation. Indeed, the physical and chemical properties of such structures vary both horizontally and vertically within an animal, and all of these properties will, in turn, be seen to change over time. For example, the tough and distensible body cuticle of the cuckoo larva of Chrysididae, in due course, is replaced by the extremely brittle and indistensible sclerotized plates of the adult wasp. Over time, cuticles vary in state of hydration, crystallinity, stiffness and strength, toughness and brittleness as well as in electrical, thermal and optical properties.

Such cuticles superimposed on structural variation are far from being physiologically stable or inert. Electrical, permeability and metabolic properties of cuticles are subject to periodic modification by the epidermis. The metabolic changes, especially with reference to metamorphosis, are enormous. Likewise, the myriad properties of the integument can be affected by the general health or sex of the animal, for example, physogastry, as well as in response to feeding, the cone-nosed sucking bug *Rhodnius prolixus* being a classical example of the latter. The structure of the cuticle can also change in response to injury, infection by parasites and parasitoids, and even daily changes in weather.

Cuticles exhibit a temporal and spatial spectrum of properties. Nevertheless, areas of cuticles can be grouped on functional grounds into the so-called "solid" cuticle typical of sclerites, apodemes, phragmata, mandibles and claws; those which consist of nothing more than resilin and covering epithelium such as the pleuro-subalar muscle tendon of the dragon fly *Aeshna cyanea*; the transitional cuticles such as found in the clypeolabral spring and distal part of abdominal tergites in the grasshopper *Schistocerca gregaria*; and, finally the thin arthrodial membranous cuticle of all insects as well as the cuticles of caterpillars and most other holometabolous larvae.

Classification Schemes

The bulk of insect cuticle is very largely occupied by procuticle and consequently considerations of possible subdivisions of procuticle have tended to assume predominance in classification schemes for the insect integument proposed through the years. This is not at all surprising since the all important epicuticle has remained intractable for so long, and because there have been relatively few extensive studies of the epidermis in conjunction with the cuticle as such. In the literature of this century there have been three more or less distinct ways of classifying the enormous variation in cuticle first extensively catalogued by light microscopists in the 19th century.

These schemes can reasonably be described in terms of the investigative approaches used in studying cuticles. For example, there is a "structural" and a "histological" way of describing integuments. Each method describes different aspects of cuticular composition and arrangement of constituents.

The structural school of classification gradually developed in the last century and is well represented by the classical work of Wigglesworth, V. which distinguished cuticle into chitinous and non-chitinous regions and many others in more recent years. Where definitions of terminology are not given in papers dealing with cuticle, it is usually safe to infer that the authors have a "structural" view of the cuticle in mind. The characteristics of cuticles seen in this way have been described by Neville, A.. In this view, exocuticles are often coloured brown or black on tanning and the chitin microfibrils of this region are arranged helicoidally and of a smaller pitch than are the microfibrils of the endocuticle. The chitin microfibrils of exocuticle are said to be more highly crystalline and better separated from the protein matrix than are the microfibrils of the post-ecdysial endocuticle. The endocuticle, apart from the above, consists of well-defined lamellae which are digested and resorbed during moulting. The various permutations of chitin architecture and the occurrence of growth layers are said to occur in the endocuticle. In this scheme, mesocuticle is regarded as impregnated with lipid but not tanned, and as a transitory stage through which presumptive exocuticle passes.

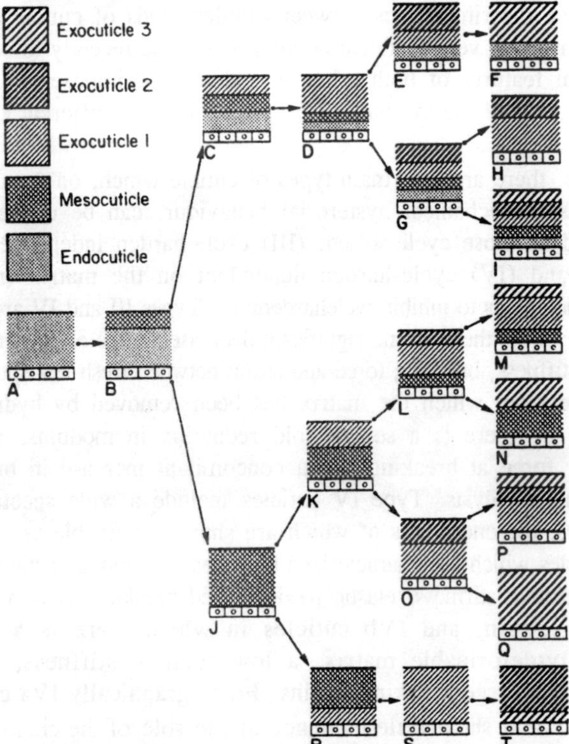

Fig. 2.18. A. Richards's scheme for the various routes and stages associated with the histological differentiation of the chitin-bearing procuticles of insects.

The "histological" scheme for subdividing procuticle is based on the colour reactions (or lack of reaction) of integumental material to Mallory's Triple Connective Tissue stains and was extensively developed by Richards, A.. Viewing cuticles in this way, one is given a qualitative indication of the extent of tanning . and hence indirectly something of the mechanical properties depending on the presence of histologically defined regions. One is also given an indication of the sclerotization route (not biochemical pathway), and the status of cuticle for a given insect can be determined with reference to the life cycle of the whole anatomical region of the animal.

Both the structural approach — which mainly describes the arrangements of chitin microfibrils in procuticle and the histological view — which mainly describes the extent to which the cuticle is tanned in a qualitative way and so protein matrix — provide much information about a cuticle, but because they are both qualitative in nature fine discriminations between similar kinds of cuticle are not readily made. A very tentative attempt was made recently to combine the useful features of both schemes and to superimpose upon them measurements of the mechanical properties of very different kinds of cuticle.

Thus, there are four main types of cuticle which, on the basis of their tensile mechanical hysteresial behaviour, can be conveniently described as those cycle-soften; (III) cycle-harden independently of matrix; and (IV) cycle-harden dependent on the matrix (matrix-dependence tends to inhibit cyclehardening). Types III and IV are easily separable: in III there are no significant decreases in the elastic modulus, relative stiffness, breaking force and strain between fresh test specimens and those from which the matrix has been removed by hydrolysis, while in IV there is a several-fold reduction in modulus, relative stiffness, force at breaking and a concomitant increase in breaking strain on hydrolysis. Type IV cuticles include a wide spectrum of properties, the end-points of which are sharply definable as follows: IVa cuticles which are characterized by a heavily tanned brittle matrix, a high relative stiffness, elastic modulus and breakin stress and low breaking strain; and IVb cuticles in which there is a highly plastically-deformable matrix, a low relative stiffness, elastic modulus but large breaking strains. Fractographically IVa cuticles are brittle and show little evidence of the role of the chitin phase, while IVb forms commonly display the influence of the fibre phase.

The majority of cuticles fall somewhere in between the end-points

Table 2.8: Classification of Arthropod Cuticles according to Mechanical, Histological and Structural Criteria

Specimen	*Proposedclassification*	*Histological classification*	*Structural classification*
Thoracic tendon of dragonfly, *Aeshna*	I	mesocuticle	none
Intersegmental membrane of locust abdomen, *Locusta*	II	endocuticle	preferred
Abdominal terga of silkworm, *Bombyx*	III	endocuticle	helicoidal
Carapace of *Limulus*(a) outer fraction	IVa	exocuticle '2' & 3	helicoidal
(b) inner fraction	IVa	exocuticle I	preferred
Wing membrane of butterfly, *Danaus*	IVb	?	helicoidal
Gena of locust, *Locusta*	(IVa > IVb)	exocuticle endocuticle	helicoidal/ preferred
Larval abdominal terga of *Agenius Zebra*	(IVa < IVb)	endocuticle	helicoidal
Elytron of adult *Agenius zebra*	(IVa > IVb)	exocuticle, endocuticle	pseudo orthogonal
Normal abdominal larval terga of *Rhodnius*	(IVa > IVb)	exocuticle, endocuticle	pseudo orthogonal
Plasticized abdominal larval terga of *Rhodnius*	III	endocuticle?	pseudo
Teneral adult, locust femur, *Locusta*	(IVa < IVb)	endocuticle	orthogonalhelicoidal/
6-hr-old locust femur, *Locusta*	(IVa > IVb)	exocuticle, endocuticle	preferredhelicoidal/ preferred
1-month-old locust femur, *Locusta*	(IVa• IVb)	exocuticle, mesocuticle, endocuticle	helicoidal/ preferred

of IVa and IVb, so that blends of these two sub-types and the analogous to case-hardened engineering materials. In destructive tensile tests of such mater outer hardened (tanned) region behaves a like a IVa material, while the inner, relatively untanned region behaves like a IVb material differences are readily apparent on fractographic examination of such a material. As the ratio two sub-types varies in a given cuticle, so does the extent of material compromise. Thus, when a relatively large fraction of brittle to plastic occurs, the simple notation IVa > IVb can be used, and conversely, IVa < IVb when there is a larger fraction in the reverse direction. In such a relative scale more or less equal fractional amounts can be denoted simply as IVa ≅ IVb. Similarly, in a IVblike structure where the matrix is not sufficiently stabilized to check the expression of chitin fibre arrangement, surface planar isotropy (*) or anisotropy (t) can be indicated by superscription, for example, IVa* > IVb* for an isotropic specimen such as locust femoral cuticle which has a brittle outer layer and an isotropic plastically deformable inner layer. Several examples have been classified in this way, and their synonymies with other classifications are given in Table. This table also shows that thehistological classification is useful in qualitatively distinguishing cuticles of types III and IV; that is, those which show well-defined differences in sclerotization, while the structural classification provides a limited amount of information on the probable planar isotropy of a given piece of cuticle.

The classification of cuticles on the basis of mechanical properties as suggested here appears to overcome many of the limitations of other schemes. For example, in type IVb cuticles, actual chitin architecture is not so important as is the fact that the absence of a stabilized matrix allows participation in the deformation process (whether of intra-stadial growth or mechanical hysteresis) of chitin-chitin cross-links and the formation of mechanical crosslinks while, in IVa cuticles, increased matrix stabilization inhibits the potential contribution of these cross-links. This scheme allows for the fact that cuticles are not only metabolically active but are also mechanically dynamic.

Whatever usefulness this latter approach may or may not have in approximating the functional properties of different kinds of cuticles, its general application in cuticular studies seems to be greatly restricted by the extremely specialized nature of experimental measurements required to study the cuticles.

Composite Cuticles

General Properties of Solid Cuticles

The term "solid cuticle" was first proposed by Jensen, M. and Weis-Fogh, T. to describe sclerotized and hardened pieces of insect exoskeleton. This term has the convenience of an engineering description in that some functional quality is implied *vis-a-vis* a term such as "sclerite", which has special meaning in morphology.

These kinds of cuticles break in a brittle manner and the fracture faces of broken specimens reveal that the cuticles are mechanically dominated by a rigid and heavily tanned protein matrix. At the other end of the spectrum are those cuticles, which we shall call "plastic cuticles", in which the proteinaceous matrix exhibits plastic flow when specimens of the material are stretched to breaking. These materials are of relatively low stiffness, low elastic modulus and breaking stress, and show a relatively large breaking strain. These sorts of cuticles demonstrate the importance of the chitin microfibrils to the mechanical integrity of the whole structure when fractured specimens are examined.

Solid cuticles can exhibit either planar isotropy or planar anisotropy, depending upon the spatial arrangement of chitin microfibrils and the extent to which the embedding medium, i.e. the protein matrix, is tanned. By way of example, stiff-brittle solid cuticle (as determined by physical measurement) occurs in very many beetle elytra, thoracic pleura of locusts, and the tanned abdominal sclerites of honeybees. Plastic cuticles occupy the other end of the spectrum and can be illustrated by the abdominal tergites and sternites of dragonflies and the membranous regions of butterfly wings.

The great majority of solid cuticles are intermediate between these two extremities, mainly because they are differentiated into exocuticle and endocuticle. In other words, stiff and brittle cuticles closely correspond to what in the histological classification scheme is called "exocuticle", and the plastic cuticles correspond to "endocuticle".

Changes in solid cuticle can be seen if one traces the physical, chemical and histological developmental events over time. Such studies are extremely few in number and information is available from a single paurometabolous insect, the locust *Locusta migratoria migratorioides,* and a single holometabolous one, the African honeybee *Apis mellifera adansonii.*

The relative changes in the amounts of NaOH non-extractable protein, stiffness (elastic modulus), and reactions to histological stains as a function of time during the development of the third abdominal tergite of the pharate and post-pharate adult honeybee and similar parameters for the larval and adult locust hind femur. It is obvious that the trends are the same in both cases. Stiffness is highly correlated with NaOH non-extractable protein so that as the amount of non-extractable protein increases (relative to total mass) so the stiffness increases. The introduction of additional cross-links in the matrix renders it increasingly insoluble (and the recovery of cross-link derivatives following drastic hydrolysis is correlated with the amounts of non-extractable protein). These relationships hold despite the fact that different tanning systems occur in the examples. The locust is β-sclerotized and the honeybee mainly quinone-tanned.

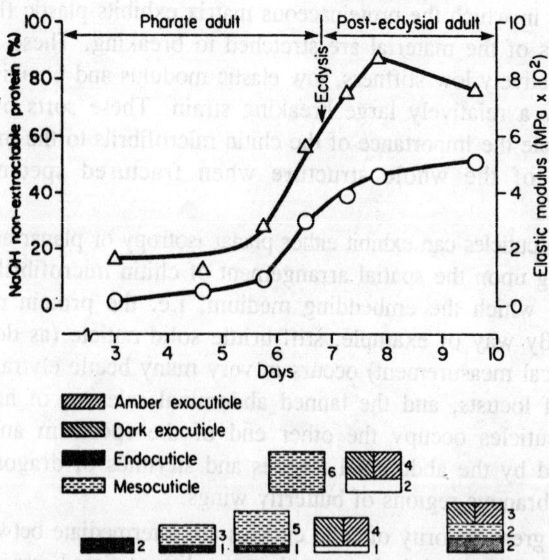

Fig. 2.19. Changes in the amounts of NaOH non-extractable protein (circles) and elastic modulus (triangles) during pharate and post-ecdysial development of the third abdominal tergite of the honeybee.

Therefore the generalization emerges that the precise nature of cross-link formation is not so important as is the actual extent of cross-linking.

That the fully formed adult locust femur and honeybee abdominal tergite are less stiff than earlier stages is only superficially anomalous because of the deposition of *more* endocuticle in both locust and honeybee *after* the exocuticle has been tanned, and the fact that the stiffness depends upon the *total* cross-sectional area of the specimen.

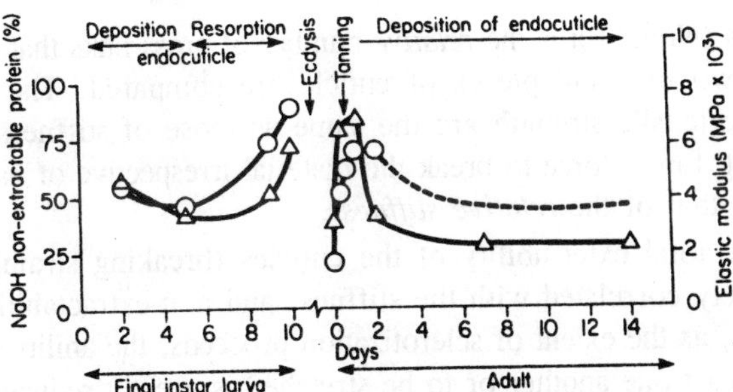

Fig. 2.20. Changes in NaOH non-extractable protein (circles) and elastic modulus (triangles) in the hind leg femoral cuticle of the locust, Locusta migratoria, that occur during larval and adult development.

If, however, one compensates for thickness and replots that data as *relative* stiffness the appropriate "intuitively" correct results are obtained. The relative stiffness is also highly correlated with NaOH non-extractable protein. A dip in the real stiffness curves will occur when the rate of sclerotization falls below that of endocuticle deposition (as in the 1-week-old adult locust and post-ecdysial honeybee). However, it is only the *relative* amount of non-extractable protein that has

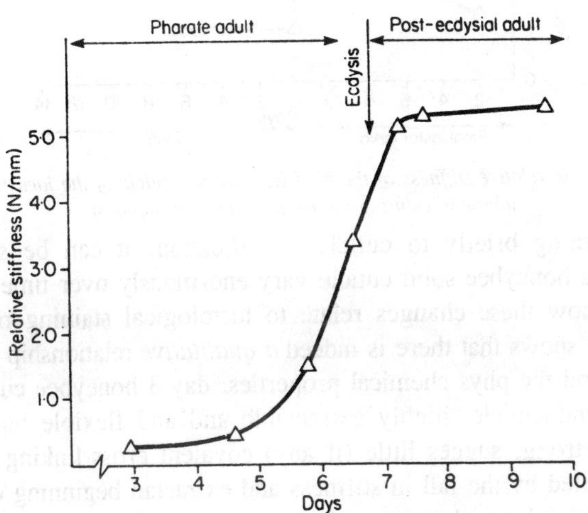

Fig. 2.21. The relative stiffness of the cuticle of the third abdominal tergite of the honeybee, Apis mellifera adansonii, *during pharate and post-ecdysial development.*

decreased in this case since the *absolute* amount cannot decrease. The *relative* stiffness (Nmm^{-1}) is directly related to the absolute amount of insoluble (tanned) matrix, hence *total* number of cross-links. For real stiffness (Nmm^{-2}) it is the *relative* number of cross-links that is reflected when two different pieces of cuticle are compared. The trends for ultimate tensile strength are the same as those of stiffness while the breaking force (force to break the material irrespective of its thickness) follows that of the *relative stiffness.*

The total extensibility of the cuticles (breaking strain) is highly negatively correlated with the stiffness and non-extractability. This is because, as the extent of sclerotization proceeds, the ability of proteins to slip past one another or to be stretched is greatly reduced owing to the cross-linking of the matrix. Where the matrix is not cross-linked to any extent the material can flow plastically; where it is well-tanned it fails in a brittle manner.

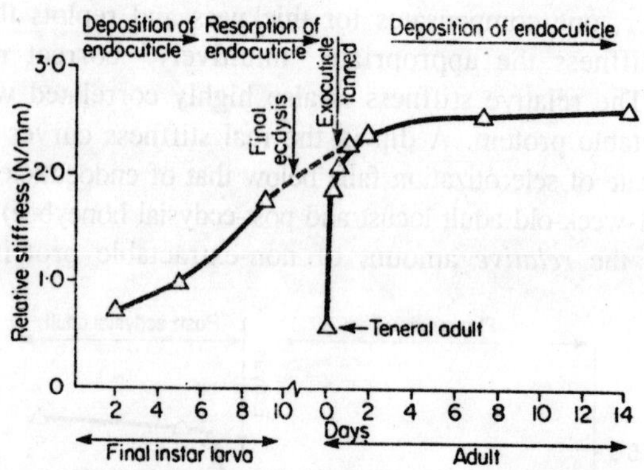

Fig. 2.22. The relative stiffness of the hind leg femoral cuticle of the locust, Locusta migratoria *during larval and adult development.*

Returning briefly to cuticle classification, it can be seen that locust and honeybee solid cuticle vary enormously over time, and we can ask how these changes relate to histological staining reactions. Figure 25 shows that there is indeed *a qualitative* relationship between staining and the phys chemical properties: day 3 honeybee cuticle is b staining endocuticle, highly extractable and and flexible but neither stiff nor strong, sugges little (if any) covalent cross-linking. This is fur supported by the fall in stiffness and extractab beginning on day 8 when blue endocuticle app again. At 64 days the cuticle is entirely red-stair mesocuticle and its extractability has decreased stiffness has increased (cross-links have 1 introduced). The appearance of non-stai exocuticle leads to further rises in stiffness and the less extractability, so that it appears that m cuticle does represent an intermediate stage on its way to becoming exocuticle. Note, however, histological exocuticle can vary greatly in stiffness so that although a qualit, relationship exists between stiffness, non-extractability and staining reactions, molecular ratior for the relationship remain equivocal. Some of the physical properties of solid cuticles are summa.

Functional Composites

The extensive research on insect cuticle civet last 50 years has been very largely dominated by fractionation studies of cuticular material into various components and the subsequent detailed analysis of very small parts, for example the study of lipids or X-ray diffraction of the chitin micelle. While a great deal has been learned in this way about minute aspects of structure and composition and sometimes, function, the literature clearly reflects the fact that cuticle biologists have been unable to see the forest for the trees. A major exception is to be seen in the approach and in the experiments of Torkel Weis-Fogh, who tried to resolve some of the difficulties associated with the energetics of flapping flight by detailed analyses of the mechanical properties of cuticle.

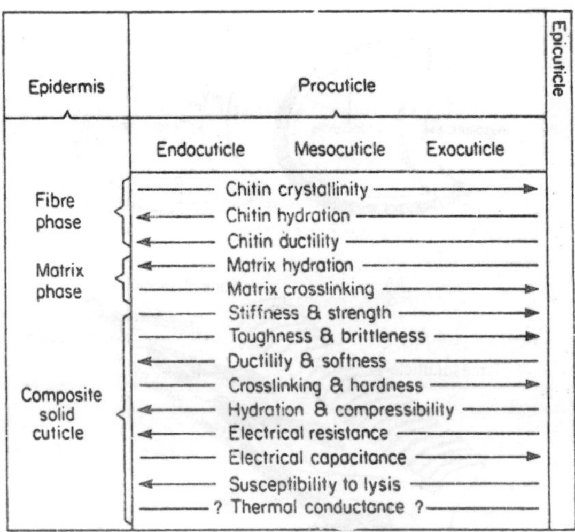

Fig. 2.23. A summary of the direction of change in some o physical properties of insect procuticle during the cour differentiation and sclerotization.

Weis-Fogh's studies of elasticity in arthropod locomotion were briefly introduced in 1958 and subsequently developed in two major classical papers on the functional properties of whole cuticles. The 1960 work established that the elastic properties of the pre-alar arm and wing-hinge elements of the locusts, *Schistocerca,* and a flightrelated tendon of the dragonfly, *Aeshna,* derive from the protein rubber, resilin. Before proceeding to further analyses of insect flight, Jensen, M. and Weis-Fogh, T. proposed a working hypothesis for cuticle consisting of a laminate of four components (excluding epidermis and epicuticle):

(1) a system of chitin lamellae parallel to the surface, not necessarily continuous but of considerable tensile strength;

(2) a rubber-like proteinaceous glue between lamellae and cross-linked to differing extents but poor in tyrosine and not tanned;

(3) additional protein which when tanned is rich in tyrosine and resin-like in nature; and finally,

(4) water.

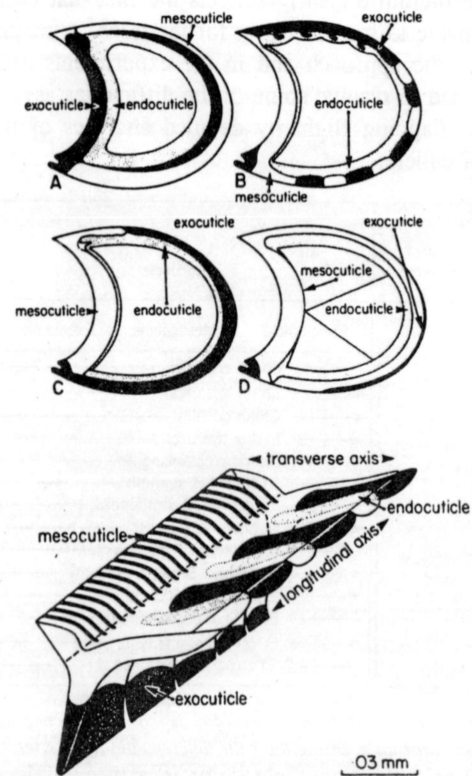

Fig. 2.24. Distribution of kinds of cuticle in the transverse planes of the galeae of moths and butterflies. (A) Agrotis exclamationis; (B) Deilephila elpenor; (C) Danaus plexippus; (D) Pieris brassicae. Distribution of cuticle types in the longitudinal axis of a portion of the galea of the moth Deilephila elpenor.

They introduced the additional constraint that the laminae are of greater than molecular dimensions to avoid the kind of molecular interlocking proposed by Fraenkel, G. and Rudall, K. which would result in very hard and brittle materials. The working hypothesis of Jensen and Weis-Fogh has proved, and continues to prove, useful as a conceptual framework for cuticular analysis. The arrangements of chitin microfibrils have in general been found to be parallel to the surface plane of the cuticle and of considerable tensile strength. The tanned protein matrix has been very well established as has the importance of water. The possible presence of a resilin-like protein in hard cuticles remains to be systematically investigated.

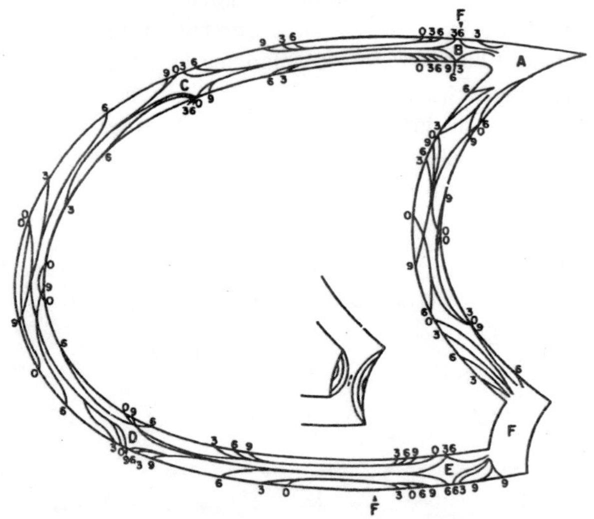

Fig. 2.25. Distribution map of the isoclinic and isochromatic fringes obtained from a compression-loaded Perspex model of a butterfly or moth galeal cross-section.

Attempts to interpret cuticular structure and modification in terms of function have been remarkably few during the last 20 years; however, a few examples point to a very modest beginning in this direction pioneered by Weis-Fogh and his colleagues. Hepburn, H. reported some simple investigations on the mechanisms which might allow for the extension and the recoil of the proboscis of moths and butterflies. Aside from important contributions from the investing musculature and internal galeal hydrostatic pressure to proboscis manoeuvrability previously established by Eastham, L. and Eassa, Y., the problem of how structure of the cuticle related to required functions of the proboscis remained.

Hepburn and colleagues approached the problem by coupling histological observations of cuticular types found in galeal cross-sections with results of photoelastic analyses of Perspex models of the galeae. Serial sections of proboscides showed variations in the kinds of cuticle present transversely and longitudinally. While the cuticle types were then defined histologically as exocuticle, mesocuticle and endocuticle, we now know, by extrapolation from subsequent mechanical measurements on cuticles, that these correspond to very stiff and brittle material, intermediate, and soft cuticles respectively. In addition, a resilin-based spring in the dorsal bar was identified.

The photoelastic analyses involved the mechanical loading of Perspex models of galeal cross-sections in a manner to simulate muscle contractions. The isoclinic and isochromatic fringes seen in the specimens placed between crossed polarizing lenses that were rotated through fixed angles were then recorded. One can then extrapolate the model data back to the histologically observed cross-sections and atterr to rationalize the cuticle types present in terms functional requirements of the coiling or uncoili proboscis.

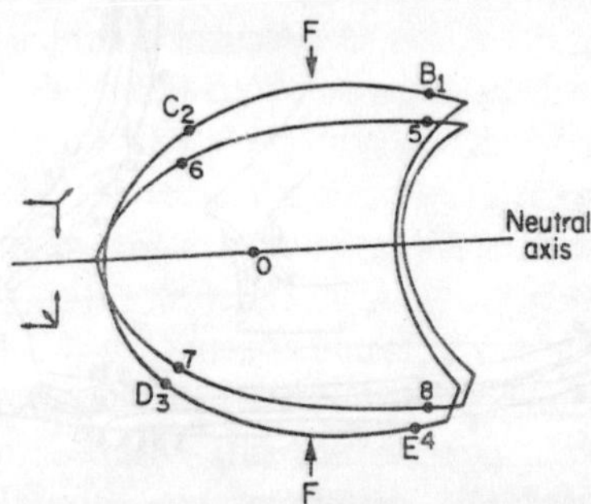

Fig. 2.26. Relative changes in position (strain) observed b different regions of a Perspex model of a galeal cross-sectior.

Functionally, these four areas must absorb transfer fluctuating compressive and tensile stress In the model the stresses are fixed at a particu loading; but in the real galea the stresses fluctuate the proboscis coils or uncoils. The regions A and F are subjected to maximum shear stress due to beiig, and produce the isochromatics typical of the kind of stress. When the models fatigue, breaks ways occur in these two regions, usually at F.

By drawing the lumen of the model, at rest a while stressed in the loading frame, and superimposing the images, the physical displacement of dividual points anywhere in the model can be determined in reference to the neutral axis, an intermediate layer which is neither compressed I stretched when the proboscis bends. By marking areas of principal stress, the behaviour of these four regions could be shown. The ar which are subjected to greatest anteroposterior compressive displacement are the CD axis (2(and the BE axis (15%). Lateral displacement a (the angle 2-0-6) = 12°, B (angle 1-0-5) = 6°. (angle 4-0-8) = 6°, and D (angle 3-0-7) = 0°. Movements are in the directions of the lines indicated above and below the neutral axis. The areas of principal stress are the same as those obtained directly by distortion. The third axis, which corresponds to force directions in the sagittal plane, is not considered for the two-dimensional model.

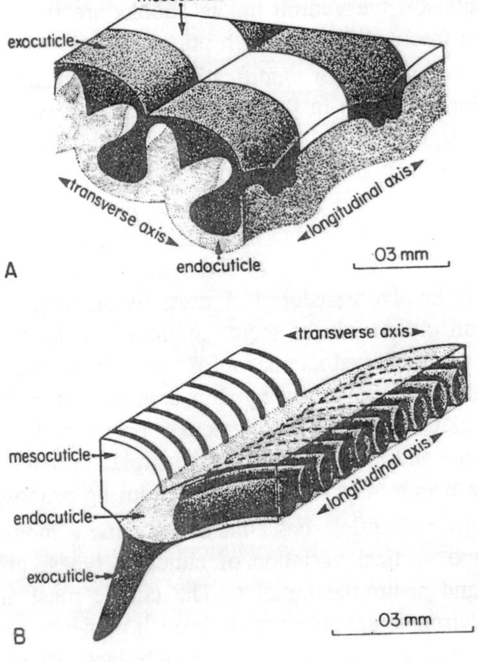

Fig. 2.27. A: Reconstruction from serial sections of the outer galeal wall of the moth, Deilephila elpenor *showing reinforcing rods of exocuticle surrounded by softer and distensible patches of endocuticle.*

The deformations just described accompany extension of the proboscis in the sagittal plane. The amount of local cross-sectional change is the lineal distance between the origin and insertion points of the individual galeal muscles. This distance decreases from the base to the apex of the proboscis while muscle declination increases.

Uncoiling is a summation of many local cross-sectional effects. In applying these model findings to the prototype, specific cuticular modifications are seen to coincide with the principal stresses and the regions of maximum shear stress.

The regions of maximum shear stress are treated in the following ways. The ventral linkage is reinforced mesally by the consistently heavily sclerotized and rigid exocuticular linkage hook. The transition of cuticle layers on the lumen side from exocuticle to either endocuticle or mesocuticle results in a flexible cuticular padding against shear in this direction. At the dorsal linkage where shear is less, there is usually a transition of cuticle types through the shear zone from a harder, less flexible to a softer, more flexible kind of cuticle. In a working proboscis the ventral linkage hooks are fixed and rigid; the dorsal hooks freely slide over each other.

There are a variety of modifications in the galea coinciding with areas of principal stress in the models. Region B is the main site of resilin. The figures show that the internal ribs of the outer galeal wall terminate just above this point, thus allowing for flexibility between the rib bases and the dorsal linkage hooks. The opposite ventral point (E) is variously modified by thinning out the exocuticle, folding it, indenting it, or eliminating it. The reactions to stress at regions C and D usually involve transferral of stress by thinning or thickening of the cuticle on either side of them. Sphingid moths have solved the problem by a complicated alternation of triangular wedges and pyriform lobes of exocuticle connected by thin exocuticular rods. These rods are supported by a softer mesocuticular padding. In the butterflies *Pieris brassicae* and *Papilio demoleus* the stress at C and E is displaced by two septa which may act as guy-ropes during proboscis extension.

The arthropod cuticle has considerably more modifications than horizontal and vertical variation of cuticular types, grossly seen as sclerotized and arthrodial cuticles. The cuticle must also allow the ingress of environmental information through numerous sensory devices such as chemoreceptors (many of which lack an epicuticle) and mechanoreceptors. These latter organs have been the subject of relatively intense research in the last few years, with attention having been paid to specific problems such as how deformation of the cuticle can be exploited by the mechanoreceptor, how the receptor cuticle is modified for function, and how deformation of cuticle leads to nervous discharge from the input signal to the central nervous system.

Barth and colleagues have investigated cuticles and Perspex models of mechanoreceptors, the socalled lyriform organs of spiders, and coupled these observations to experiments on the stimulusconducting properties of cuticular slit sense organs.

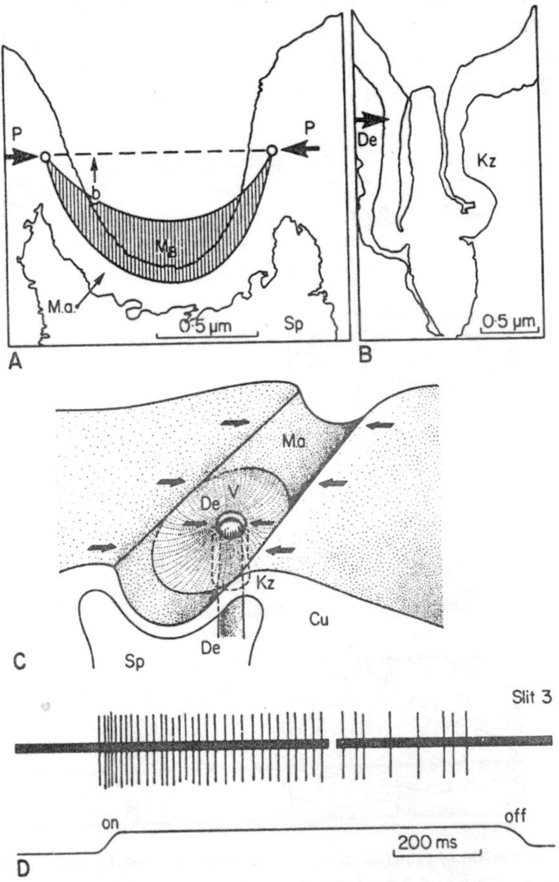

Fig. 2.28. Mechanoreceptor of a lyriform slit sense organ of the spider Cupiennius salei.

He constructed Perspex discs of different degrees of similarity to slits naturally occurring in spiders, and loaded these models in different modes (tension, compression and shear). These experiments showed that deformation of the slits in the models is greatest when the applied load is normal to the long axis of the slits.

The sensory slit in the spider is actually troughlike and if the Perspex model is slightly modified to simulate the cuticular thickening around the natural slit then any deformations related to stresses parallel to the slit are reduced, and those normal to the axis of the slit enhanced. Likewise, deformation of a slit is greatest in the middle region ana decreases at both ends. The extent of deformability, given the same load, increases with slit length. Extending these model observations back to cuticular structure, Barth observed that the lamellae of the exocuticle are arranged very closely to stress trajectories observable around 2 simple notch in a bar that is uniformly loaded it tension.

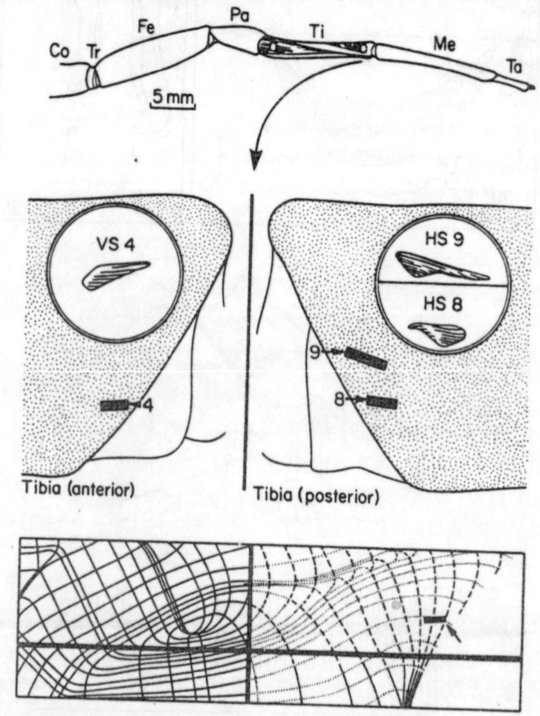

Fig. 2.29. Distribution of principal stresses as a trajectory m< of spider's leg segment. - - - - pressure, tension, • • • no stress, a lyriform organ indicated by arrow.

These observations were subsequently extend(in a far more elaborate series of experiments usil seven slits in the Perspex disc, thus simulating mo exactly the real structure found in the spider legs. In this study it was discovered that the sing slit is inadequate as a model and that a collection slits behaves differently. For example, the extent deformation of a given slit varies both with s length and whether or not the slit occupies peripheral position within the group. Moreovt different areas of the same slit vary in the extent which they deform in response to direction of an applied load. Indeed Barth found that in most cases the receptor dendrite attaches to the cuticle at the point where the greatest deformation occurs with the least force.

In extensions of this work, Barth examined photoelastic models of tibiae and showed that the topographical features of the leg coincide with pressure stress distribution in the models so that compressive forces act perpendicular to both model and to real spider. Clearly, the structure of the cuticle is interpretable in terms of model stresses. The physiological significance is that the architectural arrangement of lyriform slits in the tibial cuticle allows for a large range of stimuli to be perceived at a specific receptor site (and a marked stimulus intensity fractionation by different slits in the same mechanoreceptor group).

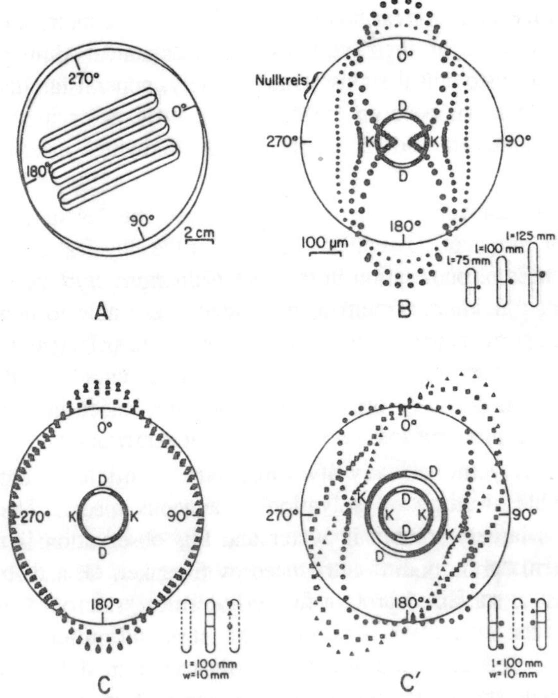

Fig. 2.30. Slit deformation behaviour of mechanoreceptor model.

Studies of the kind conducted by Hepburn, H. on the lepidopteran proboscis and Barth on the lyriform slit sense organs of spiders essentially employed optical techniques and model analysis to rationalize cuticle structure in terms of function. Much earlier work, notably that of Pringle, J. on the gyroscopic mechanism of the fly haltere and other works in that tradition, were mainly concerned with elucidating function. The recent trend has been to exploit what is known of the physical (mechanical) and structural properties of cuticle and how function depends on and derives from them. Thus, over the last few years considerable advances have been made in the understanding of how structure, hence physical properties, relate to function in the campaniform sensilla of the cricket, *Gryllus bimaculatus,* by Gnatzy, and his colleagues and in the cockroach and in the apodemes of locusts.

Arthrodial Membrane and Caterpillar-like Cuticles

We refer to "arthrodial membranes" as a general term to denote those soft and highly flexible regions of cuticle around which sclerites articulate, setae deflect in sockets, and body segments move (thus avoiding use of the term "intersegmental membrane" which has restricted meaning in morphology). These kinds of cuticles are grouped together because of their very evident mechanical similarities and reactions to histological stains. Unfortunately, general familiarity with these materials has been on the decline since freeze-drying techniques have supplanted the old-school method of "blowing" caterpillars and maggots.

The groundwork for understanding these cuticles stems from the studies of Fraenkel, G. and Rudall, K. on the physical and chemical consequences of pupariation in the flies *Calliphora erythrocephala* and *Sarcophaga faculata*. Fraenkel and Rudall were able to demonstrate several important aspects of structure here: the principal being that there is relatively little crystalline order in the cuticles and that the chitin crystallites can be oriented when suitably loaded in compression or tension, a property lost on completion of pupariation.

Similarly, the already hydrated larval cuticle is capable of considerable swelling on solvation in aqueous media. The protein fraction is largely soluble in water and this observation is the basis for the term "arthropodin" introduced by Fraenkel, G. and Rudall, K. in contrast to stabilized protein formerly termed "sclerotin" by Pryor, M.. Thus the interaction between chitin fibre and protein matrix is a very loose one and is dramatically illustrated in deformation and fractographic studies of such cuticles. Taken to destruction in a tensile manner, these cuticles are visco-plastic. There is a "necking" of the specimens as in mild steel and the material literally flows in response to continued extension.

Arthrodial and caterpillar-like cuticles are materials of very low stiffness, great distensibility and reasonable toughness. There is, however, remarkable variation in the extent to which such cuticles can be extended. A most exceptional example is the arthrodial membrane cuticle in the abdomen of the female locust, *Locusta migratoria.* This material is capable of nearly 2000% distension which is nearly three times greater than can be obtained from crossJ3 silks. Part of the explanation certainly lies in the fact that the chitin fibres are arranged normal to the longitudinal axis of the animal, and one can presume that the protein matrix is not completely bound in itself or to the chitin fraction.

The relatively large distensibility of caterpillar cuticle seems to have been exploited in intra-stadial growth by these animals. In such cases the epicuticle primarily unfolds rather than stretches but the underlying procuticle is slowly stretched in the course of a stadium, as evidenced by the continuous thinning of the lamellae (effected by hydrostatic pressure?) in the procuticle of the silkworm *Bombyx mori*. Under the appropriate circumstances, the abdominal cuticle can also behave much like arthrodial cuticle.

Arthrodial membranous cuticles, be they the general integumental exoskeleton of fly larvae or of caterpillars or proper intersegmental cuticle, have been studied by Rudall, K. and found to be basically indistinguishable on the basis of the X-ray diffraction characteristics. Despite the fact that these kinds of cuticles are *a sine qua non* of arthropody, they remain very little studied and require extensive further investigation. It is interesting to note that serious study of tanning by Pryor, M. began in the 1939-45 war, when there was considerable applied interest in the stability of laminated aircraft structures. In much the same way, perhaps the physics of plastic fabrics now used in gossamer gliders can be exploited to study the very thin and very tough fabrics that make up caterpillars and arthrodial membranes.

Cuticular Processes

There is a vast literature, largely taxonomic, describing and cataloguing the distribution and kinds of invaginated and evaginated processes associated with various parts of the integument. Basic classifications of such protuberances in the past have centred round the nature of innervation supplying the process. Snodgrass, R. emphasized what he regarded as the important distinction: whether a cuticular process was served by a single or multiple innervation or whether or not innervated at all. More recently, Richards, A. and Richards, P. have attempted to bring some order to the chaos, and have proposed a classification based on the ontogenesis for such processes as follows: multicellular in which the epidermal cells underlying the process appear similar to neighbouring epidermal cells, and others for which the epidermis is specifically differentiated; unicellular processes and subcellular ones where there is more than one projection per cell.

Multicellular processes lacking a differentiated and specialized epidermis may be movable owing to an unsclerotized basal region as in claws or leg spurs, and can be extended to include eversible osmeteria of caterpillars or defensive glands in chrysomelid beetle larvae or eversible pheromonal dispersal glands such as the Nasanov glands of honeybees or those of male Mecoptera.

Multicellular processes having a well differentiated and specialized epidermis include bristles, chaetae, hairs or setae and usually function in chemosensory or tactile ways. These are the trichoid sensilla and related structures. These processes arise from three epidermal cells that are specialized: a trichogen cell which secretes the cuticle of the process, a tormogen cell responsible for formation of the setal socket and associated cuticle, and finally the third cell is differentiated into a bipolar primary sense cell with peripheral dendrites and an axonal connection to the central nervous system. While tactile setae are usually covered with solid and continuous cuticle, that of chemoreceptive setae is porous, the pores arising during secretion of the cuticle by the trichogen cell.

Cuticular processes arising from single epidermal cells lack sockets and sense cells, and the term "acanthae" has been proposed for these now known to be widespread processes. These sclerotized processes occur in the foreguts of the adults of many orders of insects, in the reproductive system of moths and as tenent hairs on the tarsi of flies or the combs of fleas. Again, such very disparate examples are, whatever their functions, put into perspective when considered in terms of the ontogenetic origin: processes arising from a single epidermal cell lacking innervation.

Finally there are microtrichia which are very small hairs, and there are several to many of these arising as projections from a single epidermal cell. These are widespread on the surfaces of insects, particularly on the abdomen.

Peritrophic Membranes

Peritrophic membranes are secreted by the midgut epithelium, consist of chitin and protein, and occur in most insects, immature and adult, whatever their food preferences might be, solid or liquid. In addition to a role in alimentation, they may also be used in the special construction of beetle cocoons. Whether peritrophic membranes are examples of "true" integument or not simply depends upon definition.

Two general types of peritrophic membranes (PM) were described by Wigglesworth, V.: multilayered PM secreted by the surface of the midgut and single-layered PM arising from the anterior end of the midgut. This classification has become refined during intervening years with application of electron microscopy, so that Peters, W. recognizes three types based on the arrangement of chitin fibres within a PM: a reticulate or orthogonal arrangement, a hexagonal system, and finally a random arrangement of chitin fibres. If a random arrangement does in fact occur it is probably unique in insects where there are no other known cases of randomly arranged fibres in the integumentary system.

MOULTING, APOLYSIS AND ECDYSIS

General

The terms "moulting" and "ecdysis" have often been used interchangeably since the time of Snodgrass, and additional confusion was introduced when Snodgrass, R. restricted the term "moulting" to mean the separation of the old cuticle from a new cuticle formed beneath it, and held "ecdysis" to be the actual shedding of the presumptive exuvial cuticle. Modifications to these usages have been introduced by various authors during the ensuing decades.

Over the last 20 years or so use of varying terminology to describe the events associated with metamorphosis has culminated in a somewhat bitter but very edifying controversy between Wigglesworth, V. and Hinton, H.. The two essays in question are well worth the reading, if for no other reason, because they point to extreme variations in both the timing of events as well as the actual sequences of structural change that occur during metamorphosis. For example, one can encounter in the very same animal regions which are in the larval stage and yet others in the pupal stage, as in the honeybee.

In the present chapter we will follow the refinements suggested by Zacharuk, R. as follows. "Moulting" shall refer to all of the events and processes leading up to a shedding of the old cuticle.

"Apolysis" means the separation of the epidermis from the old cuticle. As such, apolysis is but one aspect of moulting. Finally, "ecdysis" refers to the shedding of the cuticle. Ecdysis is mediated by the so-called eclosion hormone, currently thought to be of universal occurrence and controlling all ecdyses in insects. "Ecdysis" would appear to be synonymous with the term "eclosion".

Added to the above, we have the terms "moult", meaning the time period and events occurring from apolysis to ecdysis; "post-moult" from ecdysis to completion of bulk tanning; and, "intermoult" from the end of the post-moult to the next apolysis.

Events of Metamorphosis

Mindful of the very large number of exceptions and special or peculiar cases, one can nevertheless describe a general outline of the events associated with moulting and ecdysis in insects as follows:

(1) apolysis along a usually anteroposterior gradient;

(2) mitotic division of the epidermal cells as well as the expansion of epidermal cell surface area and volume;

(3) the secretion of moulting fluid;

(4) formation of the outer epicuticle of the newly forming pharate cuticle and surface determination of same;

(5) secretion and formation of the new inner epicuticle of the forming pharate cuticle;

(6) activation of the enzymes of moulting fluid and the lysis and presumed reabsorption of the endocuticle of the old cuticle;

(7) deposition of the presumptive pharate exocuticle;

(8) ecdysis;

(9) expansion of the newly ecdysed cuticle;

(10) onset of tanning (which sometimes occurs before ecdysis);

(11) secretion of the endocuticle;

(12) secretion of wax;

(13) continuous deposition and partial tanning of newly formed endocuticle;

(14) formation of an apolysial membrane for the next moult where one occurs.

These are the kinds of events that occur in insects. Not all of them need occur in any one animal, nor need the sequence be exactly the same from animal to animal. Apolysis appears to be mediated, at least in part, by ecdysteroids or juvenile hormone.

While much has been made of developmental processes proceeding along anteroposterior gradients and while it is readily observable in many insects chosen at random that such gradients are very real, it has nevertheless been reported that, in the moth *Ephestia kuhniella,* moulting (= apolysis?) commences in the thorax and extends thence to head and abdomen. Coupled with the most recent discovery that eclosion hormone may come from a different anatomical source in different developmental stages of the same animal, detailed comparative developmental studies would be highly desirable.

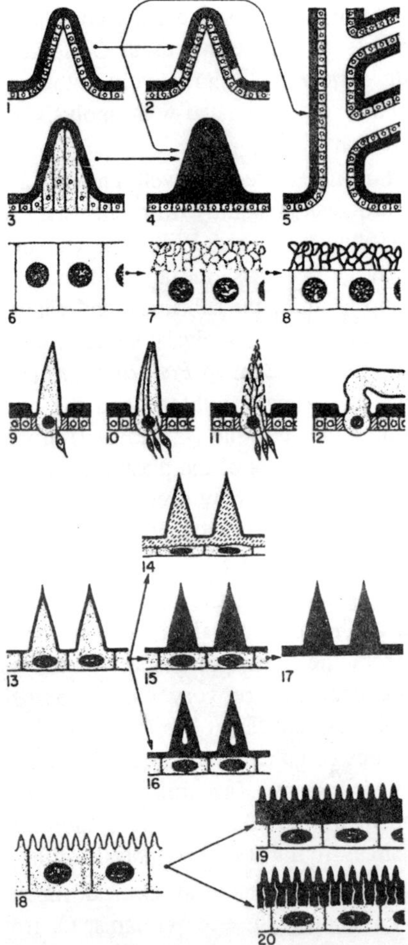

Fig. 1.31. Diagrams of multicellular projections. Cuticle solid black, cytoplasm stippled. 1 : A simple multicellular spine. 2 : Modification of a simple spine with membraneous ring at base, 3, 4: Unusual solid spine termed Brunner's organ, 5: Projection such as an antenna. 6–8: Developmental sequence for formation of the "felt chambner" of larval siracular discs of higher Diptera. 9–21: Major types of protruding trichoid sensilla. Tormogen cells are lined rather than stippled. 13-17: Sequences for development of acanthae—14: Cell processes withdrqwn from lumens of acanthae and space filled with acid mucopolysaccharide (cross-hatched) as in Panorpa; 15: cell processes withdrawn, lumens filled with cuticel as in "cornuti" in reproduction system of noths; 16: cell processes cut off by secretion of underlying cuticle as in proventiculi of certain staphylinid beetles; 17: cell processes withdrawn, lumens filled and under-laid with cuticle, then secreting cells phagocytized by haemocytes as in tenent hairs of flies. 18: Early stage with cytoplasmic filaments extending into each microtichium. 19: Solidfied microtichia underlaid by cuticle. 20: Microtichia with relationship to pore canals.

Cell division of the epidermis is usually correlated with the onset of apolysis in insects as in the waxmoth *Galleria mellonella* and cockroach *Blattella germanica*. Alternatively mitotic activity may not be at all an apparent event associated with apolysis, as suggested for some Diptera by Richards, A..

Apolysis itself is not extremely well known, and at present it seems that it may progress in slightly different ways. In their study of the silkworm, Passonneau, J. and Williams, C. suggested that the forming apolysial space occurs at the level of the subcuticle. Dense apolysial droplets appear under the subcuticle and within the endocuticle just prior to the formation of the apolysial space. Alternatively, Noble-Nesbitt, J. reported the secretion in *Podura aquatica* of a foam-like substance between the plasma membrane and the old cuticle and in the process so forming the apolysial (= ecdysial) droplets which enter the apolysial space by exocytosis of the plasma membrane. The release of apolysial droplets apparently may precede apolysis in some cases, for example *Drosophila melanogaster* or the butterfly *Calpodes* or follow apolysis as in *Hyalophora cecropia.*

There is general agreement that the apolysial space becomes larger after the secretion of the moulting fluid, and that part of this larger space is derived from the partial disgestion of the old endocuticle. Since the apolysial droplets are very closely associated with the breakdown of the old endocuticle, it has long been inferred that these droplets are collections of cuticulolytic enzymes. While exact studies of apolysis are in their infancy, Zacharuk, R. has emphasized that the physical separation of old cuticle from the epidermis (apolysis) is a process entirely separate from that of enzymolysis of the old endocuticle.

The very thin apolysial (= ecdysial, exuvial, moulting) membrane first described by Passonneau, J. and Williams, C. from *H. cecropia is* appressed to, but not connected with, the newly forming pharate cuticle and seems to be present in most kinds of insects. However, recent ultrastructural studies of certain Diptera indicate that some flies lack the apolysial membrane. The combined evidence to date indicates that the apolysial membrane is derived from the innermost lamellae of the old endocuticle because such membranes are always associated with cuticles in which there is enzymolysis of the endocuticle, and are absent from cuticles where there is no significant lysis of endocuticle.

The stability of the apolysial membrane has been credited to the occurrence of lipoprotein complexes added to the cuticle prior to apolysis in the beetle *Tenebrio molitor.* In any event, the physiological significance of the apolysial membrane remains an enigma. It occurs in some insects and is absent from others. Following apolysis and the secretion of the enzymes responsible for lysis of the endocuticle into the apolysial space, the old cuticle may be substantially digested or apparently not at all. The fate of those portions of endocuticle which are lysed is unknown. It could be that such materials are recycled in new cuticular synthesis but unequivocal supporting evidence is still wanting.

Decades of study have so often made the cuticle seem to be a physically inert and intractable substance that it is gratifying to record information as to the intense cycles of activity associated with renewal of the insect skeleton, work on the butterfly *Calpodes* particularly associated with Locke and colleagues.

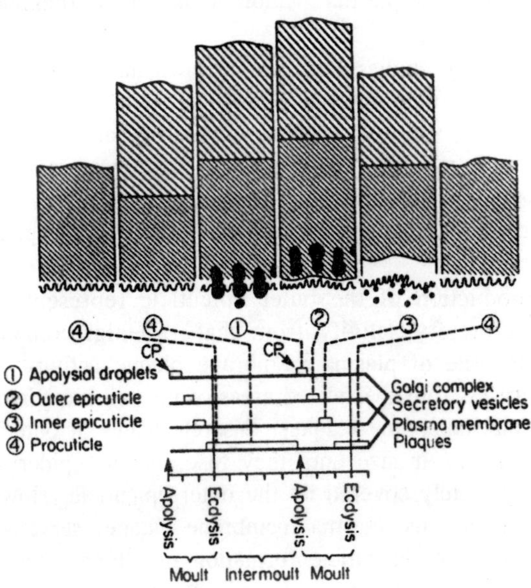

Fig. 2.32. Schematic representation of the events associated with apolysis and ecdysis in insects. There is an alternation in the activities of the plasma membrane plaques and the Golgi complex.

In this insect the first indication of moulting is the production and release of apolysial (= ecdysial) droplets which become dispersed amongst endocuticular lamellae. Prior to lysis of the old cuticle the epidermis switches over to the secretion of the outer epicuticle. Once the outer epicuticle has become tanned the proenzymes of the apolysial droplets are activated and enzymolysis of the old endocuticle commences. The epidermis then switches to the production of the inner epicuticle. Finally the epidermis switches to the production of a procuticle and to the stabilization of cuticle through various tanning pathways.

Rather than present a pastiche of fragmentary observations from various species, we summarize M. Locke's elegant essay in which the secretion and synthesis of the numerous components of the pharate cuticle are rationalized as the alternation of two different forms of epidermal secretory activity. These concern the plasma membrane plaques associated with the elaboration of the outer epicuticle and procuticle on the one hand, and the Golgi complex whose secretory vesicles are associated with the inner epicuticle and the apolysial droplets. The Golgi complex serves the additional roles of enzyme supply to multivesicular bodies and these contribute to the cyclical renewal of the apical plasma membrane and other components of the cuticle.

In a series of studies Locke has shown that although numerous proteins of the cuticle may have very different functions (inner and outer epicuticles, apolysial droplets, phenolases and peroxidases), the origins of these proteins, whenever they have been investigated, are derived from complex vesicles of the Golgi apparatus. The formation and appearance of the apolysial droplets indicates the beginning of the moult period.

The production of the outer epicuticle represents a shift in epidermal secretory activity from that of Golgi complex vesicle formation to one of plasma membrane plaque activity. The outer epicuticle arises from specialized areas of the cell membrane, the so-called plasma membrane plaques. Discrete plates of outer epicuticle gradually increase in size until they fuse and the epidermal surface becomes completely covered by the outer epicuticle. However, it is not clear whether the plasma membrane plaques serve as physical templates for outer epicuticle formation or whether they transport precursor material to the site of chemical assembly. On completion of the outer epicuticle the surface of the cell becomes separated from the pharate cuticle, the secretory vesicles from the Golgi complex are released between microvill: to form the inner epicuticle. Finally epidermal activ ity changes once more and the plasma membrane plaques (which have remained intact but quiescent during inner epicuticle formation) come to the fore in the formation of procuticular fibrous cuticle possibly the chitin microfibrils themselves.

During the moulting process many events natur ally follow on the apolysis of a given region — the synthesis of the new and partial degradation of tht old cuticle. The fate of the old cuticle during moult ing has probably been the least studied aspect o metamorphosis and investigations of cast exuvi: are extremely rare. The extent to which the oh cuticle is digested is variable. In the larvae of thi click beetle, *Ctenicera,* about one-half of the old cuticle is degraded, including material that would histologically be defined as endocuticle and the last secreted layers of mesocuticle. The situation is somewhat different in flies. In *Drosophila* substantial amounts of procuticle are degraded while in *Lucilia cuprina* apparently none of the endocuticular lamellae are broken down completely. Similarly, the fates of the cuticle digests have not been unequivocally determined. It has been suggested that they are reabsorbed through the body surface before ecdysis or that they are drunk by the insect at ecdysis.

It would appear that there are no changes in the epicuticle of the presumptive exuvial cuticle during moulting. Other structures of direct concern involving the cuticle during moulting are the tracheae and tonofibrillae or muscle attachments. J. Noble-Nesbitt's ultrastructural studies of *Podura aquatica* indicate that tonofibrillae are resistant to moulting (apolysial) fluid enzymes and that they maintain their attachments to the old cuticle until ecdysis is reached, or nearly so.

EPIDERMIS

General

The epidermis usually forms a continuous sheet of single-layered cells beneath the cuticle as well as covering skeletal invaginations such as apodemes, fore- and hindguts and the inner surfaces of evaginated structures such as wings, gills, styli, spurs, etc. (An exception is *Oncopeltus* which has two layers of cells below the procuticle Richards, personal communication). In some insects the number of epidermal cells appears to remain constant; instead of dividing mitotically they increase in size as the whole larva grows as in cyclorrhaphan flies and in many Hymenoptera. In yet other insects, such as the nematoceran flies, the epidermal cells both divide mitotically with larval growth but also increase in size. Although the epidermis is differentiated tissue, it nevertheless performs many of the kinds of functions common to other cells of other organ systems metabolism for selfmaintenance, the production of organelles and chemical synthesis and secretion.

There are other interesting, if not unique, properties of the epidermis to consider. For example, the population density of cells in the epidermis is regulated through control of mitosis with possible contributions from programmed cell death. Extensive studies of polarity and pattern, mainly in the cone-nosed bug *Rhodnius prolixus*, indicate that epidermal cells recognize both their position and orientation within the segmental gradient of the animal. While the basis of this abili remains to be more fully documented, it appears partially derive from the transfer of informatii between cells that are electrochemically polariz and structurally asymmetrical with respect to jun tional membrane contacts between cells. Epidermal cells are likewise capable of site detachment and amoeboid-like movement in to repair of cuticular wounds. Other kinds of movement vary from the changing of shape in the control of lens aperture movements towards tracheoles under conditions hypoxia. More obviously, the epidermal cells contribute much to ionic homeostasis, the synthesis of waxes, greases and the numerous instances of exocrine secretions such pheromones and compounds used for defence.

The considerable versatility of epidermal cells modulated by a variety of both extrinsic and intri: sic factors. The former include the numerous humoral factors and hormones that percola through the haemocoel, the population size of cells, physiological and heals status of neighbours, wounds, disease and parasite intercellular interactions, etc. The effects of wate temperature and humidity on the cellular milieu is well known, as also are trophic factors and chang associated with metamorphosis. Intrinsic factors are similarly numerous and include structural asymmetry of the individual cells, hence variations membrane permeability. Age, health, nutrition status and ionic conditions also affect both activeness and potential responsiveness of individu epidermal cells.

Intercellular Communication

Ionic Coupling

Since the wonderful observations of H. V. Wilson that mechanically isolated sponge cells wi spontaneously coalesce, all sorts of biologists hay been concerned with the means by which anim cells "recognize" one another and communicate. I more modern times, precisely how is function) coordination achieved in any tissue? The insect epidermis is a more restricted case because it is almost invariably a single cell thick; but of great interest because of the many different functions simultaneously performed by different areas of the same tissue. For example, given apparently excess amounts of both enzymes and the appropriate substrates, different integumental regions of the locust, *Schistocerca gregaria,* produce overlying cuticles which vary in thickness, extent of tanning, and hence stiffness; or old honeybees may return to the synthesis of wax if colony demand requires such services. Similarly, Wichard, W. and Komnick, H. have shown that the epidermal cells of the gills of mayflies are specialized so that some are involved in chloride transport; others are not.

Recent electrophysiological studies by Caveney and his colleagues have opened an entirely new and very promising line of research in understanding how cells communicate. The basic observation comes from the studies of Loewenstein, W. and Kanno, Y. who showed that the epithelial cells of *Drosophila* salivary glands are electrically coupled. In extensions of this work, the following observations emerge.

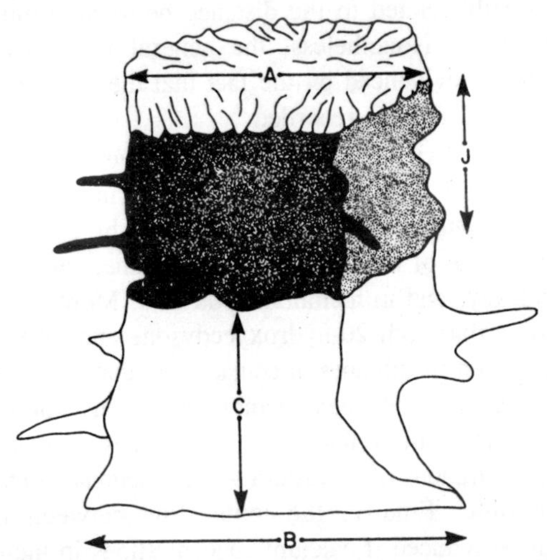

Fig. 2.33. Surface of an epidermal cell as defined by its differentiated plasma membranes. The junctional membrane area is stippled (J) while the other three areas are non-junctional and include the apical plasma membrane (A), the lateral noncontact membrane (C) and the basal membrane (B).

Each cell in the larval abdominal sternites of the beetle, *Tenebrio molitor, is* defined by a plasma membrane which has four specialized regions: an apical plasma membrane, junctional membrane, a lateral non-contact membrane and the basal plasma membrane. Adjacent epidermal cells are attached to one another via the specialized junctional membrane region of the cells. This junctional contact is further differentiated into septate and gap junctions, and it is thought that the gap junction is the most likely channel through which cells can communicate.

Caveney, S. has recently suggested that, because adjacent cells are coupled ionically when in membrane contact, the epidermis can be visualized as a functional, if not anatomical, syncytium. A further property of this arrangement is that electrical resistance between contacting regions is relatively low *vis-k-vis* non-contact regions, so that any movement of ions from one cell to another is an anisotropic one. Because of the relatively high resistance between two cells not in junctional membrane contact, there is likely to be little leakage of ions into such extracellular spaces.

The extent to which a series of cells might exert an influence is limited by the distances between them, since the magnitude of the potentials is directly related to the distance between stimulating and recording electrodes. Nonetheless, the general resistance of these pathways is low, as evidenced by the fact that the movement of ions over about 25 cells has been recorded.

In a compelling series of measurements, Caveney, S. has shown that ionic conduction very likely fluctuates during the pupal-adult metamorphosis of *Tenebrio molitor,* as evidenced by great changes in the specific resistance of non-junctional membrane, resistance of the intercellular pathway, and in membrane potential. Moreover, Hax, *W. et al.* have shown that both 20-hydroxyecdysone and cAMP increase the resistance between membranes in contact, and that juvenile hormone decreases the resistance between membrane regions not in contact. The electrical resistance of nonjunctional membranes is greatly affected by changing concentrations of potassium while junctional contact regions are less susceptible. Finally, the ionic flow between junctional membranes can be reduced if calcium concentrations in the epidermal cytoplasm are increased.

Growth and Size

An insect cuticle can change in pattern, form and size during the course of metamorphosis. Such changes are inevitably related to major events in the epidermis. Among these, there are major variations in the rates and kinds of lytic and synthetic activities, cell turnover through ageing and death and mitoses. Most of these activities are greatly affected by the distribution, kinds and amounts of hormones in a given region of the body. For example, juvenile hormone both directs the kinds of synthesis performed by certain cells as well as affects the rate of cell death. Moreover, it has also been suggested that there is an inverse relationship between the levels of moulting hormone and cAMP, and a positive correlation between bursicon activity in the haemolymph and epidermal cAMP with the implication that cAMP might mediate bursicon production and release. Against this, it might also be the case that moulting hormone regulates cAMP levels by depressing them, and bursicon by elevating them. So, it is conceivable that bursicon might mediate cAMP production and not vice-versa. This problem remains unresolved.

Cell turnover would seem to be a very pertinent but unstudied aspect of growth and size. Since the ultimate size of any instar must eventually depend upon the amount of material synthesized by the underlying epidermis, the ratio of functional to non-functional cells ought to be considered. Such information would be of considerable interest. Another problem attending the study of growth is how to measure growth in insects. Growth of the exoskeleton (increase in surface area) can be restricted to postecdysial expansion prior to hardening or can be intrastadial in nature — intussusceptive or creep. Mitotic activity in the epidermis is initially associated with synthesis of the entire dorsum of the pulvillus. These observations are quite encouraging as they hold the possibility of locating the precise region of the genome responsible for the synthesis of individual constituents of the cuticle.

BASEMENT MEMBRANE

In most insects there is a layer of material immediately below and connected to the overlying epidermis. This layer of the integument is undoubtedly the one about which we know least. On the basis of electron micrographs it would appear that the basement membrane consists of collagen fibres in *Locusta migratoria* and in a beetle, *Hydrocyrius columbiae*. The origins and formation of the basement membrane in *Rhodnius prolixus* have been studied electron microscopically by Wigglesworth, V. who concluded that it derives from plasmatocytes in the haemolymph.

CONCLUDING REMARKS

The incredible variation to be found in the integuments of insects ultimately derives from simple principles. Extracting these principles from the integument is, of course, our main task. As I see the literature on the insect integument, there are many interesting and important problems to be resolved. For example, the general applicability of the Bouligand model to insect cuticles is in doubt and more work is needed here. We know nothing of the molecular conformations of cuticular proteins and such knowledge is essential to an understanding of the physical chemistry of tanning. We know virtually nothing of the significance of the basement membrane or of procuticular lipids. We do not know to what extent formation of cuticle is a phenomenon of crystallization or of cell-directed synthesis and assembly. While exciting discoveries have been made concerning the electrophysiology of the epidermis, this work is in an embryonic phase and sorely in need of greater efforts. Is piezoelectric transmission a viable route by which information can enter the epidermis? How is resilin synthesized? Those concerned with insect pest management will want to know the mode of action of insecticides inhibiting chitin synthesis. What we have learned to date holds more for titillation rather than education of the intellect.

3

TANNING

Tanning agents are the compounds which during sclerotization are secreted by the epidermal cells into the cuticle, where they are acted upon by enzymes to make them reactive and able to connect to proteins and other components. According to the definition compounds deposited unchanged in the cuticle are not to be considered as tanning agents, although they may influence the properties of the structure. This could be the case for uric acid and 3,4-dihydroxyphenyl acetic acid, which can be extracted unmodified from cuticle of some Coleoptera.

N-acetyldopamine (I) is so far the only compound for which there is convincing evidence that it functions as a tanning agent for insect cuticles. Other phenolic compounds are used for tanning of non-cuticular materials, such as 3,4-dihydroxybenzoic acid (IV) and 3,4-dihydroxybenzyl alcohol (V) for tanning of cockroach egg-cases. Gentisic acid (VI) and 2-amino-3hydroxybenzoic acid (VII) are used for stabilizing the silks of the silk moths *Antheraea pernyi* and *Hyalophora cecropia,* respectively. Even these species, which use unconventional phenolic compounds for tanning the secreted materials, apparently use Nacetyldopamine for stabilizing their cuticle.

Characterizing cuticular enzymes is difficult; it is not possible to determine the specific activity of the enzymes as long as they are inside the cuticular structure, and there is considerable risk of modifying the enzymes when they are solubilized. Often it can be a problem to get enzyme activity out of the cuticle, where it may be bound to the

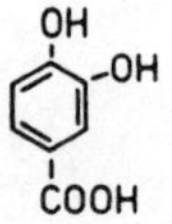

IV

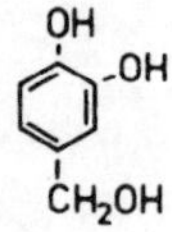

V

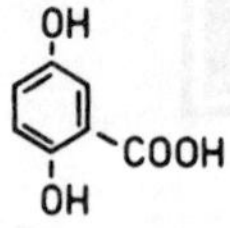

VI

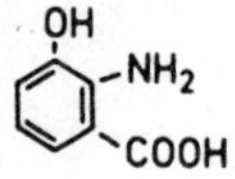

VII

cuticular structure in some unknown way. Extracts from whole insects are of little use when studying the cuticular enzymes, since insect hemolymph is rich in prophenoloxidase, which can easily be transformed into the active enzyme. The activation can be caused by limited proteolysis, or by lipid-like substances. In *Drosophila* it has been shown that activation of the proenzyme is the result of a cascade of reactions, which is started when animals are damaged. The activated enzymes from insect hemolymph tend to aggregate readily and can easily be adsorbed to cellular debris or to fragments of cuticle. Only investigations, using isolated and purified cuticles, can therefore be assumed to give reliable information on the enzymes involved in sclerotization. That the cuticular diphenoloxidases are not identical with those from hemolymph was shown by Hackman, R. and Goldberg, M. who, from dipteran larval cuticle, obtained extracts containing both mono and diphenoloxidase activity, and the activities differed in several respects from the activities obtained from hemolymph. Ohnishi, E. showed that *Drosophila virilis* larval cuticle contains an inextractable enzyme catalyzing the oxidation of various diphenols and of dimethyl-p-phenylenediamine, which is not a substrate for the hemolymph enzyme. The enzyme has been characterized in more detail by Yamazaki, H., who used powdered, acetone-dried cuticle from fruit-fly white puparia. The cuticular powder has a wide substrate specificity. It will oxidize *ortho*- and *para*diphenols and aromatic diamines, but has negligible activity towards tyrosine. The activity is unusually thermostable, and it is not inhibited by a number of compounds, such as thiourea, which inhibits the hemolymph enzyme. The activity resisted all attempts at solubilization, but a similar enzymatic activity could be obtained in soluble form by trypsin digestion of cuticle from silkwork pupae *(Bombyx mori),* collected during larval-pupal ecdysis, when the cuticle is still unhardened. As soon as the cuticle has become sclerotized it is impossible to obtain the enzyme by trypsin treatment, indicating that it becomes sclerotized together with the structural proteins in the cuticle. There is a drastic increase in enzymatic activity in the intact cuticle during the first hours after ecdysis, indicating either activation of a proenzyme or deposition of fresh enzyme. The enzyme activity drops again to reach a low level at about I day after ecdysis. The oxidase released by trypsin treatment was purified, and a nearly homogeneous preparation was obtained and partly characterized. It was concluded that presumably it is a copper protein, and it was classified as a laccase. Enzymes resembling that obtained by Yamazaki from silkworm pupal cuticle have since been obtained from other sources, such as unsclerotized locust cuticle and blowfly larval cuticle. If the ability to catalyze the oxidation of para-diphenols is used as criterion for the presence of laccases, this group of enzymes is of widespread occurrence in cuticles of insects (unpublished observations).

Laccases have been purified after trypsin digestion of unsclerotized cuticle from the locusts, *Schistocerca gregaria* and *Locusta migratoria.* From blowfly *(Calliphora vicina)* larval cuticle two different laccases have been obtained both with and without trypsin treatment. Since some insect cuticles may contain proteolytic activity, the possibility cannot be ruled out that the enzyme is released by proteolysis from blowfly larval cuticle, even when no external trypsin is added. From *Drosophila* larval cuticle (wandering stage) small amounts of oxidase activity have been extracted without addition of trypsin (unpublished observation). The substrate specificity corresponds to that of the other enzymes mentioned, but it has not been obtained in sufficient quantities for purification and characterization.

Electrophoresis of purified preparations in polyacrylamide gels under denaturing conditions gave three bands having molecular weights between 62,000 and 70,000 in the case of silkworm pupal cuticle, and two bands with molecular weights between 90,000 and 100,000 from S. *gregaria* cuticle. The major laccase-like enzyme from blowfly larval cuticle gave only a single band with molecular weight about 90,000. The multiple bands obtained from two species could be due to limited proteolysis during release of the enzymes by trypsin digestion, but it may also indicate the presence of isoenzymes.

The insect laccases have a rather broad substrate specificity, as a number of substituted catechols and hydroquinonesare readily oxidized, whereas neither mono-phenols nor meta-diphenols are oxidized. Compounds carrying negative groups on the side-chain, such as protocatechuic acid and dopa, are rather poor substrates, whereas a positive charge has little effect on the rate of oxidation. *N*acetyldopamine is a good substrate for the enzymes, although not the best. The *K*. value for *N*acetyldopamine was determined to be 0.53 mM for the blowfly larval enzyme and 1.3 mM for the enzyme from locust cuticle. The enzymes are inhibited by cyanide, fluoride, and azide ions, whereas complexing agents such as o-phenanthroline, dipyridyl, and phenylthiourea have little effect.

The pH optima of the laccases are in the acidic range, varying from about pH 4.5 for the major blowfly enzyme to pH 5.5 for the silkworm enzyme. It is doubtful whether the cuticular pH will ever be as low as this, and such low pH optima need not have a functional importance. There is a tendency for the pH-optima to be slightly higher for intact cuticle than for the solubilized enzymes.

There is no evidence that diphenoloxidases other than those described are present in locust or silkworm pupal cuticle. From the larval blowfly cuticle we have obtained two diphenoloxidases different from the major laccase enzyme described above. One is also of the laccase type, but has a pH optimum at a slightly alkaline reaction and differs from the major laccase in its substrate specificity.

The other enzyme is not a laccase, but resembles the hemolymph diphenoloxidase so much that it seems likely that it is the same enzyme which occurs both in cuticle and in hemolymph. The enzyme belongs to the tyrosinase type, as it catalyzes the hydroxylation of tyrosine to dopa and the oxidation of o-diphenols to o-quinones. It has no activity towards p-diphenols, nor towards aromatic diamines. The enzyme is strongly inhibited by thiourea and phenylthiourea, it has a pH optimum near neutrality, and is less thermostable than the cuticular enzymes of the laccase type. The enzyme is apparently strongly aggregated in the extracts, since it is excluded from 5% polyacrylamide during electrophoresis. Such aggregation has been analyzed in more detail for the hemolymph diphenoloxidase, where it has been shown that the enzyme is present as an unaggregated, inactive precursor, which after activation tends to form large, ill-defined aggregates. There is some evidence that the cuticular tyrosinase-type enzyme in blowfly larvae also occurs as an inactive precursor, which is activated by proteolysis when the cuticle is damaged. At least it has been found that when pieces of blowfly cuticle are cleaned at pH 8.5, they contain high tyrosinase activity when tested soon after cleaning. If they are cleaned at pH 5 the activity is low, but increases rapidly after transferring the samples to pH 8.5. This increase can be prevented by the presence of trypsin inhibitors. The activity of the laccase-type enzymes is unaffected by the pH during preparation and by the presence of trypsin inhibitors, indicating that the activities of the various enzymes in blowfly larval cuticle are controlled in different ways.

Only a limited number of insect cuticles has been investigated with respect to the types of diphenoloxidases they contain, but it appears that the laccase type of enzyme is present in cuticles destined to be sclerotized, both when sclerotization occurs soon after ecdysis, as in locusts and silkworm pupae, and when it occurs at a later stage, as in puparium formation in Diptera. The tyrosinase type of diphenoloxidase has been found in soft cuticles which are never sclerotized, such as larval cuticle of the silkworm, *Hyalophora cecropia* (unpublished observation), and in blowfly larval cuticle, which has to function first as a soft cuticle during larval life and then as a hard, sclerotized cuticle in the puparium. The blowfly larval cuticle is the most complex example we have encountered with regard to enzymatic complexity. The correlation between presence of tyrosinase-like enzyme and soft texture of the cuticle has led to the suggestion that this type of phenoloxidase is not involved in sclerotization, but in healing of damaged cuticle. According to this suggestion the laccase type of diphenoloxidase should be the enzyme involved in sclerotization.

Peroxidases are another group of enzymes which might be involved in oxidizing diphenols to reactive intermediates. It has been suggested they may be responsible for catalyzing the cross-linking of the cuticular, rubber-like protein, resilin, by oxidizing protein-bound tyrosine residues to free radicals. Peroxidases may also be involved in sclerotization of hard, cuticular structures. Peroxidase activity has been demonstrated both in the epidermal cells and in cuticle, but the enzyme has not been purified and characterized from insects. Since catalase and other heme-containing proteins can have slight peroxidatic activity, and catalase has been found present in some cuticles, it is necessary to purify and characterize the enzyme to ascertain whether it should be classified as peroxidase or catalase. The substrates which are commonly used to demonstrate peroxidase histochemically, such as diaminobenzidine and benzidine, are also oxidized to colored products by the cuticular laccases. Accordingly a positive reaction with these substrates will not prove the presence of peroxidases.

Although it is well established that the diphenol, N-acetyldopamine, is the tanning agent in many cuticles, enzymes other than diphenoloxidases might also be involved in activating the compound. Some years ago evidence was obtained that locust cuticle contains an enzyme activity catalyzing the release of tritium specifically located on the carbon atoms of the side-chain of N-acetyldopamine, The enzyme is firmly bound to the cuticular structure, and its properties have only been characterized in intact cuticle, since it has not been possible to obtain active extracts. The relationship between this' activity and the diphenoloxidase present in cuticle is ambiguous, since it has only been possible to compare the two activities in intact pieces of cuticle, and such comparisons are made complicated by penetration and diffusion problems. The comparison showed that the activity, which is responsible for the release of tritium from the side-chain of N-acetyldopamine resembles the laccase activity. The two activities have the same high thermostability, the same stability, nearly the same pH optimum and inhibitor specificity. The small differences which were observed could be explained as diffusion artefacts due to the differences in assay conditions. It was observed that the trypsin-solubilized laccase would release tritium from the aromatic ring of labelled N-acetyldopamine, while tritium located on the side chain was not released to any significant extent. This observation can be explained by assuming that an enzyme different from the laccase is responsible for releasing side-chain-located tritium from N-acetyldopamine, when intact pieces of cuticle are used. An alternative explanation is that a single enzyme is responsible for both activities in the intact cuticle, and that trypsin digestion modifies the enzyme by limited proteolysis, thereby destroying the side-chain-directed activity, while leaving the laccase activity intact.

The one-enzyme hypothesis was supported by the observation that *ortho*- and para-diphenols, which are good substrates for the laccase, will competitively inhibit the release of tritium from the side-chain of N-acetyldopamine when intact cuticle is used as enzyme source. Moreover, the best substrates for the laccase were the most effective inhibitors of tritium release. This relationship can be expected, if it is the same active site which binds both N-acetyldopamine and the various other diphenols.

Results obtained by characterizing the reaction products, when N-acetyldopamine is used as substrate, indicate that intact cuticle contains enzyme activities which are different from the purified laccase. When intact locust cuticle acts upon N-acetyldopamine the main product is a dimeric compound for which the structure XII was first suggested. On the basis of NM studies the structure has been revised, and the correct formula is XIIIa and/or XIIIb. It is probably a mixture of isomers which is formed.

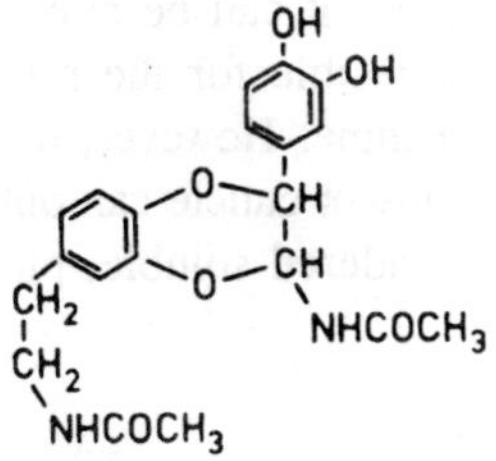

XIII a

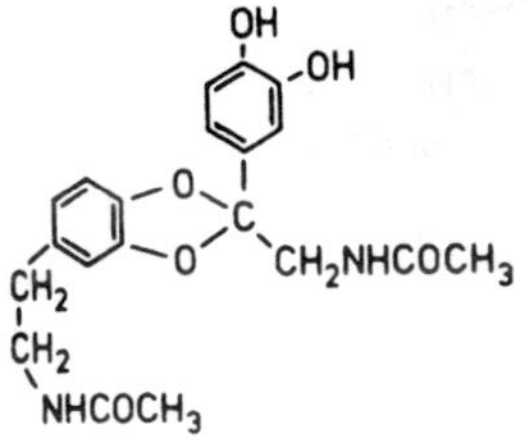

XII

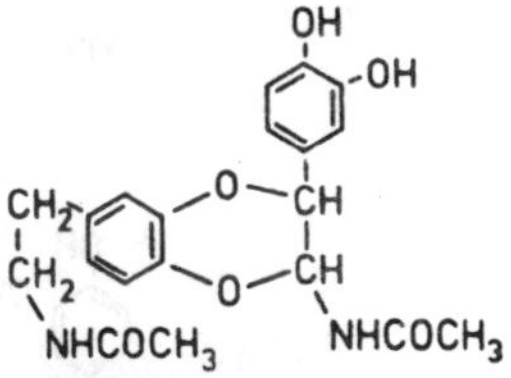

XIII b

Another dimeric compound has recently been isolated from the incubation medium after reacting N-acetyldopamine with intact locust cuticle. It is characterized by having a double bond in one of the side-chains, as shown in structure Xl. It is formed in much smaller amounts than compound II.

None of these dimeric compounds are formed when purified locust laccase is incubated with Nacetyldopamine. The fact that they are formed when intact cuticle is used as the enzyme source can be explained by assuming the presence in cuticle of an enzyme that will introduce a double bond into the side-chain of N-acetyldopamine. The presence of such an enzyme can be demonstrated by adding some amino acids or small peptides to the incubation medium. They will partly inhibit the laccase, and in their presence a metabolite of Nacetyldopamine accumulates in the medium. It has been isolated and identified as dehydro-N acetyldopamine (XV). It is a far better substrate for the laccase than N-acetyldopamine, and it is accordingly consumed as rapidly as it is formed if the laccase is not inhibited.

The "desaturase" activity, responsible for the formation of the double bond, has not yet been characterized in detail, nor has it been solubilized, but it can be assumed that it corresponds to the enzyme activity responsible for the release of tritium from the side-chain of N-acetyldopamine. However, the true identity of the various enzymatic activities in insect cuticle can only be ascertained when all the activities have been rendered soluble, purified, and characterized in detail.

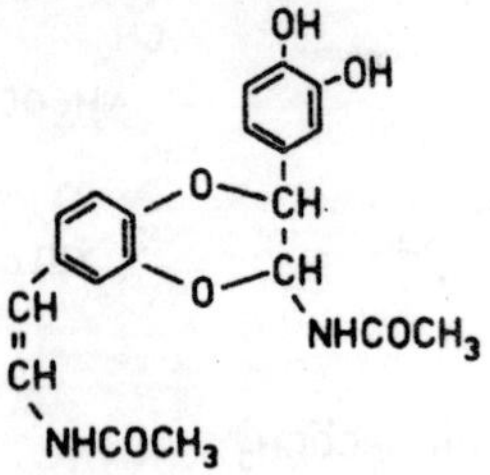

XIV

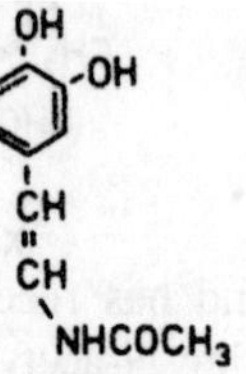

XV

THE CHEMISTRY OF TANNING

Several attempts have been made to obtain insight into the chemical reactions going on during sclerotization by studying the fragments which can be obtained after degradation of fully sclerotized cuticle. Such studies are necessary, but can be misleading, as artefacts are easily produced during degradation. Dennell, R. used alkaline hydrolysis, under reducing conditions, of blowfly puparia to liberate what he identified as a substituted aminohydroquinone. The identification was mainly based upon one-dimensional paper chromatography and color reactions without having appropriate standards available, and the results cannot be considered reliable.

Acid hydrolysis of sclerotized cuticles has been used to liberate substituted catechols, among which ketocatechols (XVI-XIX) are the dominating species. The type of ketocatechol obtained depends upon the exact conditions of hydrolysis, and it has been argued that they are degradation products of a common precursor in the cuticle. However, small amounts of ketocatechols have also been obtained after extraction of cuticle under nonhydrolytic conditions, and it was suggested that they are present as such in the cuticle before extraction, and that their function might be to act as antioxidants and not to be involved in protein stabilization via cross-linking.

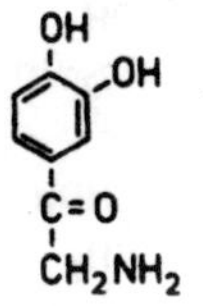

XVI (C=O, CH_2NH_2) XVII (C=O, CH_2OH)

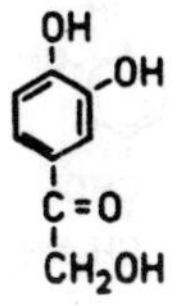

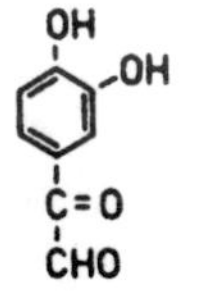

XVIII (C=O, CHO) XIX (C=O, CH_2, $NHCOCH_3$)

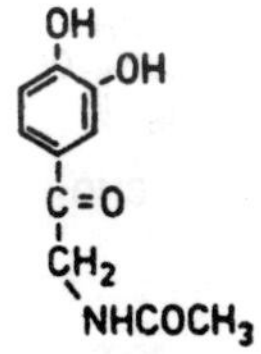

Digestion with proteolytic enzymes has also been used to liberate ketocatechols and other catechols from sclerotized cuticle. Catechols obtained in this way have not been linked to the cuticular structure, but may have been trapped inside the cuticular network. The ketocatechols, which can be obtained in this way, only account for a small fraction of what can be released by acid hydrolysis. The free ketocatechols are thus only minor components in sclerotized cuticle: they may have a function as antioxidants, or they may be unimportant byproducts formed during sclerotization. The following catechols have been obtained from proteolytic digests of locust sclerotized cuticle: dopamine (IX), N-acetyldopamine (I), arterenol (XX), N-acetylarterenol (XXI), hydroxyketocatechol (XVIII), protocatechualdehyde (XXII), protocatechuic acid (IV), and several dimeric compounds, including compounds XIII and XIV.

When exposed to acid hydrolysis the dimeric compounds are degraded, and the moiety with two free phenolic groups is recovered as a ketocatechol. A fraction of the ketocatechols obtained after acid hydrolysis of cuticle will be derived from the dimeric compounds present in the cuticle. The fraction is, however, of quantitatively minor importance since the dimer, which is present in largest amounts, can account for less than 1 % of the total amount of ketocatechols which are obtained. Higher oligomers are presumably also present in sclerotized cuticle and will give ketocatechols on acid hydrolysis, but so far we have only indirect evidence for their existence.

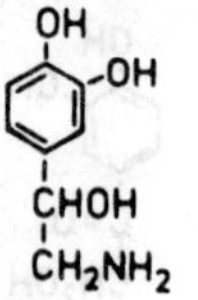

OH
-OH
CHOH
CH_2NH_2

XX

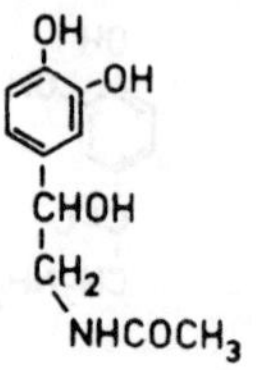

OH
-OH
CHOH
CH_2
$NHCOCH_3$

XXI

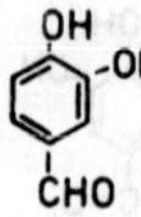

OH
-OH
CHO

XXII

During proteolytic digestion of sclerotized cuticle most of the catechol-containing material is released as a mixture of compounds which, although they are not ketocatechols, give ketocatechols by acid hydrolysis. During hydrolysis with acid a number of free amino acids are also formed. Despite several attempts to fractionate the proteolytic digest it has not been possible to obtain a pure compound for identification; neither has it been possible to separate the moiety which gives amino acids on hydrolysis from that which gives ketocatechols. It appears that the catechols are chemically linked to peptide material, but the nature of the links is still unknown. Since the same ketocatechols are obtained after acid hydrolysis of the dimeric compounds and of the catechol-containing peptide material, it can be assumed that links from both the α- and β-carbon atom in the side-chain of *N*acetyldopamine are involved in both cases. A significant fraction of the phenolic groups from tyrosine residues and of the ε-amino groups from lysine residues in the cuticular proteins is not available for substitution in sclerotized cuticle, and these groups may be involved in the linkage. The protein-linked catechols are presumably present in polymeric form, and a reliable estimate of the degree of polymerization is needed.

There are several compounds among those obtained by mild acid hydrolysis which are not easily accounted for, such as arterenol and 3,4dihydroxyphenylglycol (XXIII), catechols with a secondary alcohol group in the β-position of the side-chain. They may be reduction products from the corresponding ketones, but there is no supporting evidence that such a reduction takes place during hydrolysis. The compounds might also be derived from material linked to proteins with a single, hydrolyzable bond, as presumably the ketoderivatives are formed by hydrolyzing two such bonds. More difficult to explain is the isolation of a ketocatechol acetyl ester (XXIV) after mild acetic acid hydrolysis of sclerotized cuticle. The acetyl group is not introduced during hydrolysis, since it is present also when formic acid or propionic acid is used for hydrolysis. The compound is derived from *N*acetyldopamine incorporated into the cuticle, but how the acetamide group is transformed into an acetyl ester is so far inexplicable.

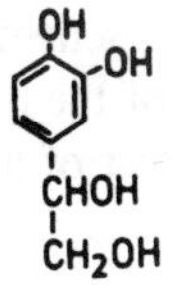

OH, OH
CHOH
CH_2OH

XXIII

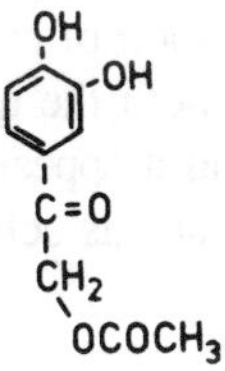

OH, OH
C=O
CH_2
$OCOCH_3$

XXIV

Ketocatechols can be obtained from nearly all sclerotized cuticles from insects. Their amounts can correspond to more than 10% of the cuticular dry weight, and values down to 0.2% have been obtained for locust cornea and intersegmental membranes. Ketocatechols can be obtained from cuticles sclerotized before ecdysis as well as from those sclerotized after ecdysis. Cuticles which only sclerotize after ecdysis will not yield ketocatechols until the process of sclerotization is initiated. Ketocatechols have not been obtained from sclerotized cuticle from non-insect arthropods or from apterygote insects (Thysanura, Collembola, Diplura) (unpublished observations), indicating that the ability to fly and the ability to produce ketocatechol-yielding material have been developed in parallel during evolution.

It was mentioned above that β-alanine plays a role in the sclerotization of some cuticles. Acid hydrolysis releases the amino acid in significant quantities from sclerotized dipteran puparia, but not from the still unsclerotized, white puparium. It has, to my knowledge, not been possible to obtain β-alanyl derivatives by limited degradation of sclerotized puparia. The bonds between β-alanyl residues and the cuticular material are very labile towards acid treatment. If the suggestion that β-alanyldopamine is a tanning agent in fruit-fly puparia is correct, and if it is incorporated by the same pathway as *N*acetyldopamine, it should be possible to obtain β–alanyl-substituted ketocatechols by very mild acid treatment of *Drosophila* puparia. We have searched for such β-alanyl derivatives without success. Degradation studies have so far been of no help in investigating the role of β-alanine in sclerotization.

PROPOSAL FOR A SCLEROTIZATION MECHANISM

A revised scheme for the chemical reactions occurring during sclerotization has recently been proposed. It is based upon the results obtained from degradation studies of sclerotized cuticle and from characterization of enzyme activities in cuticle. The scheme incorporates aspects of both the quinone tanning scheme and of the β-sclerotization scheme, and it appears to be able to account for most of the observations made on various sclerotized cuticles.

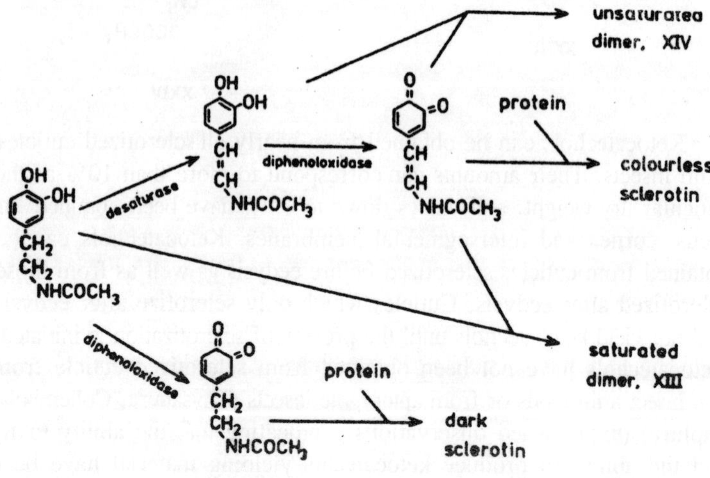

Scheme III

The scheme suggests that cuticles which are able to sclerotize contain two different enzyme activities, both of which can use N-acetyldopamine as substrate. One of the enzymes is a diphenoloxidase which, in the presence of oxygen, will oxidize N-acetyldopamine to the corresponding quinone. The enzyme is supposed to be identical with the laccases which have been isolated from various insect cuticles. The quinone can react with the cuticular proteins as assumed in the classical quinone tanning scheme, and the result is darkly colored sclerotins. The other enzyme, a desaturase, can remove hydrogen from the sidechain of N-acetyldopamine, thereby introducing a double bond in the side-chain. The unsaturated compound, dehydro-N-acetyldopamine (XV), is a substrate for the same diphenoloxidase (laccase) as that which oxidizes N-acetyldopamine to a quinone. The resulting unsaturated quinone (XXV) possesses an extended system of conjugated doublebonds, and it can be predicted to be a highly reactive molecule. It reacts with the ortho-diphenols available (N-acetyldopamine and dehydro-Nacetyldopamine) to form substituted benzodioxins (XIII and XIV) as well as with the cuticular proteins. The cuticle is thereby sclerotized, and the end-product is a colorless material.

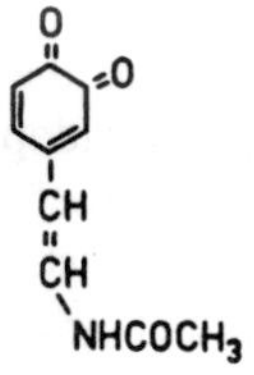

O
=O
CH
CH
$NHCOCH_3$

XXV

The main features of the scheme are:

(1) quinone tanning and β-sclerotization occur at the same time and use the same primary substrate, N-acetyldopamine;

(2) quinone tanning can occur without simultaneous β-sclerotization, whereas β-sclerotization cannot occur without some quinone tanning going on at the time;

(3) the relative importance of the two pathways depends both upon the relative activities of the two enzymes and upon the affinities of the two substrates, N-acetyldopamine and dehydro-Nacetyldopamine, for the diphenoloxidase;

(4) since the unsaturated quinone is two oxidative steps removed from N-acetyldopamine, it can react spontaneously with two groups

and there by form a cross-link, and no protein-bound intermediates, which need reoxidation, are formed as assumed in the quinone tanning scheme.

Evidence is available for all the steps suggested in the scheme, but not all the evidence is equally conclusive. The diphenoloxidase (laccase) has been purified from three different types of cuticle, and it is active *in vitro* towards both N-acetyldopamine and dehydro-N-acetyldopamine. The desaturase has not been purified and characterized, but its activity has been demonstrated in locust femur and fruit-fly puparial cuticle. Oxidation of dehydro-Nacetyldopamine *in vitro* leads to the formation of the unsaturated dimer (XIV), and when Nacetyldopamine is added to the incubation medium the saturated dimer (XIII) is also formed. If dehydro-N-acetyldopamine is oxidized by the diphenoloxidase in the presence of foreign proteins some of the substrate is coupled to the proteins, but it is not yet known what type of bonds is formed. If unsclerotized locust cuticle is incubated with dehydro-N-acetyldopamine some of the compound is incorporated into the cuticle; the cuticle remains colorless, it becomes stiffer, and the proteins become insoluble. After acid hydrolysis of the cuticle the usual types of ketocatechols can be recovered.

The most serious weakness in the scheme is that the linkages between the derivatives of *N*acetyldopamine and the proteins are unknown, despite many attempts and efforts to identify them. There is good evidence that the linkages exist, but they have proved highly elusive. The two most important problems in cuticle sclerotization today are: What is the precise nature of the interprotein links?, and How is sclerotization controlled?

CONTROL OF SCLEROTIZATION

It is necessary that sclerotization occurs in the right place, at the right time, and to the right degree. Too little or too much sclerotization might prove fatal to the insect, but it is not known how precise the control has to be; it will probably vary from species to species. It seems reasonable to assume that a precise control of the degree of sclerotization will not be essential for the dipteran puparium, as long as it is sufficient to protect the animal developing inside the puparium. As the wings of flying insects are exposed to strong and varying bending and twisting forces, they will need mechanical strength, some resilience, and extreme lightness. Too little sclerotization could make them too soft, and too much sclerotization could make them too brittle and heavy. However, little experimental evidence bearing on this problem is available, due to difficulties in obtaining reliable quantitative measurements of mechanical properties and of the degree of sclerotization of selected samples of cuticle.

No reliable methods exist for determining the total degree of sclerotization of cuticles. It would be valuable if we were able to determine the number of cross-links between macromolecules in a piece of cuticle, or to determine the total amount of diphenolic material incorporated into the cuticle. The yield of neutral ketocatechols after hydrolysis of cuticle samples in 1 M HCl can be used as an approximate measure for the amount of *N*acetyldopamine incorporated via the β-sclerotization pathway, but we cannot determine how much is incorporated via quinone tanning. Ketocatechol analyses can be used as an indication for the degree of sclerotization, when there is evidence from enzyme analyses that β-sclerotization is the dominating pathway during sclerotization.

The enzymatic characteristics of locust femur cuticle indicate that about 90% of the total sclerotization is due to 8-sclerotization, and the yield of ketocatechols probably gives a reasonably good estimate of the total sclerotization in this sample. The method has been used both to follow the progress of sclerotization and to obtain an impression of the variability in sclerotization of femur cuticle among mature individuals of *Locusta migratoria* grown together in the same cage. The ketocatechol yield varied from a minimum value of 0.44% to a maximum value of 2.30% of the femur dry weight; the mean value for 58 samples was 1.26%, and the standard deviation was 0.54%. The mean dry weight of the femur cuticles was 11.17 ± 0.98 mg, and the heaviest samples weighed 1.5 times more than the lightest (unpublished observations). The ketocatechol yield can thus vary with a factor of five, although the animals were reared together under controlled conditions. It is not known whether a similar variability is present in natural populations. Neither is it known whether the differences in yield of ketocatechols correlate with other properties of cuticle. If the ketocatechol yield is a reliable estimate of the degree of sclerotization these preliminary measurements indicate that there is no very precise control of the degree to which locust femur cuticle is sclerotized.

The amount of enzyme activity, measured as the release of tritium from the side-chain of Nacetyldopamine, has been determined in cuticle from various body parts of locusts. The values were compared to the yield of ketocatechols from corresponding parts from animals analyzed either 1 day or 10-12 days after ecdysis. No correlation was found between enzyme activity and the yield of ketocatechols, which indicates that it is not the amount of enzyme which determines to what degree the cuticle is sclerotized, and that the limiting factor must be looked for elsewhere.

Injection of dopamine, generally labelled with tritium, into adult locusts about 2-6 h after ecdysis, resulted in incorporation of much of the label into the cuticle. The distribution of label between samples of cuticle from various body parts corresponded closely to the degree of sclerotization measured as ketocatechol yield from the corresponding parts of control animals. The mandibles were the only exception; their incorporation of labelled dopamine was nearly twice that expected from the ketocatechol yield. The mandibles, in contrast to the other samples of locust cuticle, are dark brown, and they will liberate tritium both from the side-chain and from the ring positions in Nacetyldopamine, indicating that both quinone tanning and f3-sclerotization occur in this sample. The results indicate that a single dose of dopamine is distributed into various parts of cuticle corresponding to the rate at which they are sclerotized via the β-sclerotization pathway. It therefore seems likely that it will be the uptake of dopamine or N-acetyldopamine in the epidermal cells which will be the rate-limiting step, and that the epidermal cells from various parts of the integument will compete for the available dopamine. The cuticle above the epidermal cells with the greatest ability to take up dopamine/N-acetyldopamine will accordingly be the areas which will become most heavily sclerotized.

The degree of sclerotization will also depend upon the total amount of sclerotizing agent synthesized in the animal. It was mentioned above that N-acetyldopamine is mainly synthesized from tyrosine via dopa and dopamine, and that there is no general agreement on where this synthesis takes place. Both hemocytes, oenocytes, and epidermal cells have been suggested to play a role. It appears that the synthesis is under hormonal control, and there is suggestive evidence that both ecdysteroids and bursicon can be involved in initiating sclerotization. Ecdysteroids induce sclerotization in ligated abdomens of blowfly larvae, and one of the direct effects of 20-hydroxyecdysone in this system is the induction of *de novo* synthesis of the enzyme dopa-decarboxylase in the epidermal cells. To my knowledge there is no evidence that this enzyme is involved in the regulation of sclerotization in systems other than puparium formation in dipteran species.

In several instances it has been possible to induce cuticle synthesis and tanning *in vitro* by the addition of 20-hydroxyecdysone to the medium. Thus it appears that the epidermis of some insects is equipped with the complete system for synthesizing the tanning agent and can be activated by appropriate hormone treatment. Marks, E. and Sowa, B. report that cockroach epidermis can produce cuticle *in vitro,* when stimulated by 20hydroxyecdysone, although the cuticle appears incomplete and it is not sclerotized. They suggest that some controlling factor, for instance bursicon, is missing in the *in vitro* system.

Bursicon was described as a hormone necessary for cuticle tanning and endocuticle formation of flies and other insects. There is some evidence that the hormone can stimulate the uptake of tyrosine into the hemocytes and thereby initiate its conversion to N-acetyldopamine. Bursicon also appears to be able to stimulate the epidermal cells directly, as demonstrated by a conspicuous rise in epidermal cyclic AMP when the level of bursicon activity is increased in the hemolymph of *Tenebrio molitor.*

From the available evidence it can be concluded that the synthesis of the tanning agent is hormonally controlled, and that at least 20-hydroxyecdysone and bursicon may be involved. The mechanisms of the hormonal controls are poorly understood, and the possibility that further hormones are involved cannot be excluded.

The ultrastructure of insect epidermal cells has been studied in detail, and much information has been obtained on the structural changes in the cells during the molting cycle. We lack a correspondingly detailed study of the biochemical changes in the epidermis during the same period, so that the ultrastructure can be correlated with function on the molecular level.

It appears that the post-ecdysial sclerotization of insect cuticle is controlled on two levels: a central, hormonal regulation of the onset of sclerotization and a local control of the extent and type of sclerotization. In some insects parts of the cuticle are sclerotized before ecdysis, while the animal is in the pharate state, and in the honeybee it has been shown that this pre-ecdysial sclerotization utilizes the same chemical reactions as post-ecdysial sclerotization. Nothing is known about how the pre-ecdysial sclerotization is regulated, but it is possible that it is initiated by a rise in the level of ecdysteroids.

4

BIOCHEMISTRY OF CHITIN

In the course of its evolution the insect has developed a cuticle composed primarily of chitin and cross-linked protein (sclerotin) that enables it to cope with the complexities of its environment, providing for growth, mobility, protection and communication. The former component of cuticle, chitin, is the simplest of the glycosaminoglycans and is β (1 → 4) homopolymer of N-acetylglucosamine $(C_8H_{13}O_5N)_{n>>1}$. This polysaccharide also occurs in nature combined with other materials in algae, fungi and marine invertebrates. In cuticle and peritrophic membrane it is thought to exist as long microfibrils strongly bound to protein and possibly to pigments. The morphogenesis of the chitin-protein complex into cuticle is one of the many mysteries in developmental biology and can be understood adequately only by comprehending both chitin and protein assembly, whether self or directed, as well as the synthesis and catabolism of these components. This chapter is concerned with the metabolism of chitin, which is not understood in sufficient detail to justify making broad generalizations. Insects make great use of chitin in structural tissues, in which respect they differ markedly from higher animals. We think that some of these pathways are relatively unique in the animal kingdom and thus they may become targets for precisely designed biochemical pesticides. It is no doubt apparent that the potential for exploitation of chitinous materials in insects is very great.

The metabolism of chitin in micro-organisms is better understood than in other life forms. In recent years research on chitin in insects has intensified because of the serendipitous discovery of insecticides which apparently block the formation of chitin. Many laboratories are now characterizing in more precise biochemical terms how chitin is synthesized, degraded and assembled into supramolecular structures in many different organisms. Consequently, it is difficult to make a comprehensive review-of chitin metabolism which does not soon become outdated. The following is an effort to summarize the precarious state of our knowledge in this field of research involving insects.

CUTICLE STRUCTURE AND COMPOSITION

Insect cuticle is composed of chitin and protein plus several other components (lipids, salts, pigments, etc.) which contribute only a few percent to its total dry weight. Chitin can vary from as little as several percent to as high as 60% of the total dry weight. This structural polysaccharide is therefore of utmost i ortance to the growth and development of insects. Before focusing our discussion on chitin alone, it is worthwhile to review cuticle structure in general and the interaction of chitin with the other components.

Research investigating arthropod cuticle has increased in recent years, and the topic has beer approached in a variety of ways. These studies use physical, chemical, kinetic and immunologica techniques to answer important and pressing questions in molecular biology, e.g. control of metamorphosis, the mechanism of hormone induction structure/function relationships of macromolecule! and receptor sites, chromosomal puffing supramolecular assembly and insect control Reviews have been published discussing various aspects of cuticle biology, and include several com prehensive texts, one citing work done before and two citing more recent papers. Shorter reviews covering selected topics include those by Wigglesworth, V.; Hackman, R.; Locke, M.; Richards. A. and Andersen, S. Hepburn, H., discusses integument structure elsewhere in this volume.

The cuticle of insects cannot be viewed as a static structure as it changes almost continuously throughout the life of the animal. Structure and composition differ from species to species, and even differ from one part of an animal to another. An idealized overview will be given, though, to formulate a simple model. The terminology has changed over the years, making a review of the literature confusing, but the generally accepted terms arc those suggested by Weis-Fogh, T..

The basic structure of the insect integument comprises a layer of epithelial cells covered by an overlaying cuticle which the cells have secreted. The cuticle is multilayered, each layer having been sequentially deposited by the cells, the outermost being deposited first. The horizontal layers are traversed by vertical pore canals, through which cell secretions can be transferred to the surface of the cuticle.

The cuticle can be divided into several sections: the epicuticle, exocuticle, mesocuticle and endocuticle, the last three making up the procuticle. The former covers the whole of the insect except for chemoreceptors, midgut and the ends of gland cells. Less than 1,um in depth, it is fairly uniform throughout the animal. The general opinion is that the epicuticle contains no chitin although some authors have disagreed.

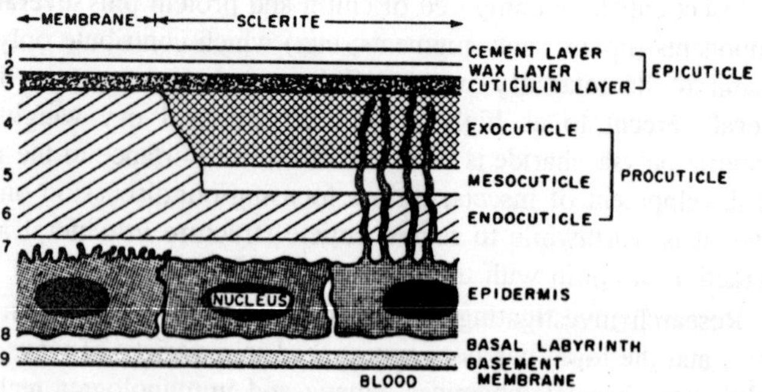

Fig. 4.1. Idealized cuticle structure. A diagrammatic section through the integument of an insect showing two types of cuticle: membrane and sclerite. The procuticle layer(s) are the chitin-containing portion of the cuticle. The horizontal layers are transversed by the pore canals which provide a route from the epidermal cells to the epicuticle.

The outermost layer of epicuticle, not always present, is the cement layer. Composed of tanned protein and lipid, its probable role is to protect the wax layer just beneath it. Studies where the wax was damaged or dissolved have demonstrated that the wax layer is essential for maintenance of the insect's water content.

The procuticle is that portion of the cuticle which contains chitin. The properties of the procuticle differ widely from one part of the animal to another and most probably depend on the particular combination of chitin and proteins present.

The exo-, endo- and mesocuticular layers were initially described on the basis of histological staining. When exposed to Mallory's triple strain the exocuticle stained blue, the mesocuticle red and the endocuticle was refractile to the stain. The exocuticle is secreted before ecdysis and appears in the electron microscope to have chitin fibrils in what has been called a "highly crystalline pattern". The proteins are tanned and this layer is not dissolved at ecdysis. The mesocuticle is less organized and untanned. The endocuticle, which is next to the epidermal cells, appears striated in electron micrographs. The striations are growth layers that are visible because of a change in the orientation of the chitin microfibrils. The endocuticle is resorbed from one instar to the next and in starvation can be used as a nutrient source.

The evolutionary selection of an exoskeleton composed of crystalline chitin and cross-linked protein has dictated a pattern of discontinuous growth for insects who must shed the exoskeleton and replace it by a larger "shell" periodically. In general terms a new cuticle partly forms within the old cuticle, the inner portion of the old cuticle is dissolved and resorbed, and then the outermost layer (exuvium) of the old cuticle is shed. In many larval insects, exuviae are consumed by the new instar so that many components are recycled.

Morphological studies during insect development have contributed to the understanding of the individual components of cuticle and their dynamics during the molting process. The chitin microfibrils as visualized by the electron microscope seem to originate directly at the epidermal cell border, reinforcing the idea that the enzyme chitin synthetase is located in the cell membrane. Chitin is known to be produced from precursors located in the epidermal cells, but the origin for cuticle proteins is less clear.

The cuticle proteins of interest here are those associated with chitin. They can be classified into two groups:

(1) those which bind chitin covalently or noncovalently and contribute to the supramolecular assembly and disassembly of cuticle; and

(2) those which are directly responsible for the synthesis and degradation of chitin.

Interactions between chitin and cuticle proteins can be weak and non-specific or strong and involve covalent linkages. For example, cuticular proteins from *Calliphora* larvae bind crab chitin noncovalently and non-specifically. There have been several reports of amino acid or protein residues in chitin preparations and some covalently bound chitin protein mucopolysaccharides have been fractionated. Poly-Nacetylglucosamine from sarcophagid puparial cases remains covalently conjugated with peptides even after extensive chemical and enzymatic degradation, as well as chromatographic separation.

The differences in solubility properties between chitin and protein sometimes make it possible to carry out their chemical characterization separately. Unlike most structural macromolecules, many of the cuticle proteins are water-soluble, at least until they become sclerotized. The following remarks will deal with the chemical properties of cuticular proteins. Chitin will be addressed in the next section.

To date the study of cuticle has not resulted in a systematic picture of the structure, synthesis or degradation of its protein components, although some generalizations can be made. The most common method of studying cuticle protein is to subject the cuticle tissue to a series of extraction buffers and then to characterize the components of each extraction, in recent years relying on data from gel electrophoresis. Investigations of the properties of proteins extracted from insect cuticle have been conducted primarily by Richards, A.; Andersen, S.; Andersen, S. and Weis-Fogh, T. and Hackman, R.; Hackman, R. and Goldberg, M.. Attempts also have been made to study proteins in different layers of cuticle separately.

Cuticular proteins tend to have low isoelectric points (pH 3-6), relatively low molecular weights (10-80 K, especially <20K) and they contain a rather high percentage of amino acids with hydrophobic side chains. More recently structural proteins from sarcophagid larval exoskeleton have been partially characterized as part of a study on cuticle sclerotization.

Lipids represent only a small percentage of the total dry weight of cuticle, but their role as a water barrier between the insect and its environment is crucial for survival. No reports of chitin having an interaction with lipid are known, although chitin-lipid complexes are indeed very possible, at least non-covalent ones. Cuticular lipids are extremely heterogeneous and do not possess the degree of similarity apparent among cuticle proteins.

CHITIN STRUCTURE AND OCCURRENCE

Chitin was first described in 1811 by Henri Beacomot, a member of the French Academy of Science. In the course of his research on mushrooms he treated specimens with warm, dilute alkali and isolated chitin, a product he termed "fungine". Odier, A. found the same substance in insect cuticle, which he renamed "chitin" from the Greek word for tunic or envelope. The chemical composition of chitin (glucosamine and acetic acid) was first described by Ledderhose, G. and confirmed by Gilson, E.. In spite of this information, the idea that chitin could be defined as a polymer remained controversial for many years. Many believed it to be a mixture of glucosamine and an unspecified nitrogen compound. Chitin was found to be a polysaccharide by Purchase, E. and Braun, C.. More recently there has been renewed interest in chitin chemistry, its commercial preparation and its laboratory uses, prompting several reviews and conferences.

Chitin is widely distributed in animals and represents the skeletal polysaccharide of several phyla, i.e. Arthropoda, Annelida, Mollusca and Coelenterata. In several groups of fungi, chitin replaces cellulose as the structural polysaccharide. In insects chitin has been found in the body wall, gut lining, cuticle and muscle attachment points in all species tested.

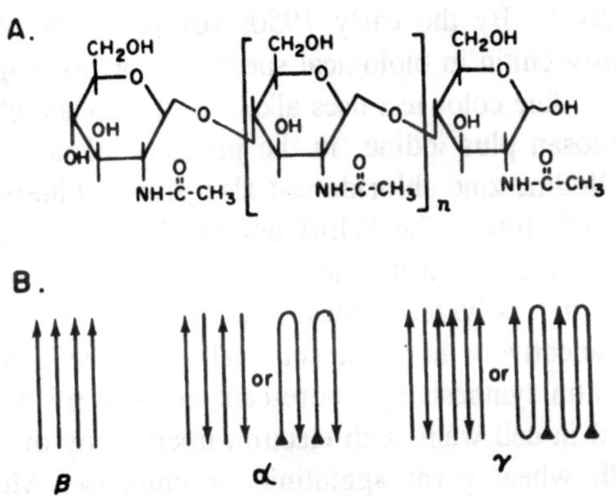

Fig. 4.2. Chitin structural components. (A) β (1 → 4) linked 2 acetamido-2-deoxy-D-glucose, the repeating chitobiose unit of a chitin chain. (B) The molecular interpretation of the arrangement of chitin chains in the unit cells of three crys tallographic types of chitin. The forms are termed, left to right: β (parallel chains), α (antiparallel chains) and γ (two chains up for every chain down). All three types are normally associated with protein.

Chitin is a polymer of N-acetylglucosamine joined in β (1 → 4) glycosidic links. Its structure may be considered a substituted cellulose with an N-acetyl moiety replacing the OH on carbon-2 found in cellulose. Because of their chemical similarities, research on chitin and cellulose has often been intertwined. The completely Nacetylated chitin polymer may only be an idealized structure, however, as studies with lobster chitin show the acetyl group is missing on some residues. The free amino group may serve as a nucleophilic attachment site for other components in cuticle. *In situ* insect chitin is present as microfibrils and is surrounded by cuticular proteins. Depending on the, type of cuticle, the chitin/protein matrix contains different relative amounts of each component. Highly elastic cuticle has relatively low levels of chitin. The fibrilar arrangement suggests that only the outermost chitin chains in any group are available for bonding with cuticle protein. Some researchers contend, however, that chitin in cuticle is saturated with protein and that non-covalent binding between the two is non-specific. The chitin microfibrils in cuticle are organized in any of several patterns which have been studied in detail with the electron microscope. The microfibril model replaces an earlier model of cuticle structure in which chitin and protein were described as monomolecular sheets occurring alternately.

Early attempts to describe and identify chitin were based on its solubility properties, i.e. it being the cuticular component insoluble in hot, dilute alkali. By the early 1950s several color reactions were used to identify chitin in biological specimens or histological sections. The chitosan/iodine color test uses alkali to deacetylate chitin, forming chitosan; chitosan plus iodine, in the presence of acid, turns violet. The diphenoliodine-zinc chloride test also gives a blue/violet color in the presence of chitin. The Schiff polysaccharide reaction has been used to indicate chitin, but this non-specific reaction cannot distinguish chitin from other polysaccharides. Calcofluor, a dye that seems to stain chitin selectively in yeast cell walls, has been used to study patterns of chitin synthesis by fluorescent microscopy. Chitin has also been localized in cell walls with electron microscopy of colloidal gold particles with wheat germ agglutinin or chitinase. More recently, fluorescein-labeled isothiocyanate conjugated with wheat germ agglutinin or with chitinase has been used to locate chitin *in situ*. Methods for estimating the amount of chitin in biological samples have been reviewed elsewhere. Hackman, R. and Goldberg, M. have reported a colorimetric assay useful for determining microgram amounts of chitin present in arthropod cuticles. The assay detects 2,5-anhydrohexoses produced from deacetylated chitin. Color is developed with 3-methyl-2-benzothiazolone hydrochloride and $FeCl_3$, and is stable for 90 min.

The preparation of chitin has been a constant problem in experimental work due to its limited solubility in aqueous systems. Chitin is soluble only in caustic solvents such as hot, concentrated mineral acids which may lead to covalent bond rearrangements. Procedures for the preparation of chitin have been published and reviewed elsewhere. The usual source for chitin is crab or shrimp shells. Most preparative procedures include long incubations in concentrated mineral acids or hydrogen peroxide solutions, exposing the sample to oxidizing conditions and to pH and temperature extremes. These harsh treatments are known to deacetylate chitin and to cause

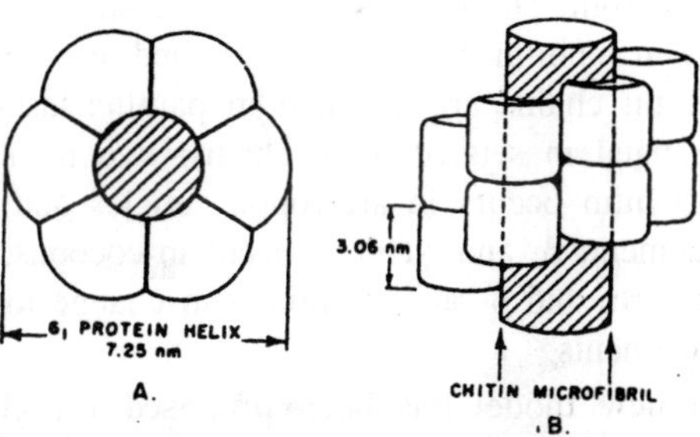

Fig. 4.3. Chitin/protein complex of α-chitin. The arrangement of protein subunits around the chitin core in a microfibril, based on X-ray fiber diagrams of ichneumon fly ovipositor (Megarhyssa). *The fibril is viewed (A) perpendicular to the fiber axis and (B) along the fiber axis. The protein subunits are represented as globular structures that repeat in 3.06 nm along the axis. Adapted from Blackwell, J. and Weih, M. and used with permission.*

fragmentation of the long chains. Colloidal chitin can be prepared by dissolving chitin in cold, concentrated hydrochloric acid and then filtering it quickly into ice cold water. This method has problems, because the product may not resemble native chitin. Oligomers of chitin have been prepared from native material for use in studying protein-saccharide interactions. Hackett, C. and Chen, K. report a method that extracts chitin from ascospores of a fungus *(Sordria brevicollis)* and from eggs of a nematode *(Ascaris lumbricoides)* at neutral pH using an ozone-peroxide-EDTA solvent. Nonaqueous solvents such as hexafluoroacetates or chloroalcohols mixed with mineral acids can also extract chitin. A chitin-protein complex was prepared from *Heliothis zea* integument by boiling in 1 sodium dodecylsulfate for 2-4h. All of these procedures can be applied to insect cuticle for the preparation of insect chitin.

Researchers have used various physical techniques to investigate the architecture of chitin polysaccharide, including electron microscopy, nuclear magnetic resonance and electron diffraction. The most successful description of the supramolecular structure, though, has come from X-ray diffraction studies. Descriptive models have been developed, but these must not be interpreted as representing chitin in general. Instead, they represent only those parts of a specimen that show a high degree of order. Examples of tissue that are especially suited to X-ray study, because of the minimal amounts of foreign material present, are diatom fibers *(Thalassiosira fluviatilis)* and cuttlefish pen *(Loligo)*.

Deproteinization of the samples sharpens the pattern somewhat, but raises questions about the "nativeness" of the specimen under study. Chitin chains are helical and contain two residues per turn. Three polymorphic forms of chitin have been described, differing in the arrangement of their polymer chains. The a-chitin structure was proposed by Carlstrom, D. and is an antiparallel sheet. The hydrogen bonding pattern in a-chitin has been determined recently. In the β-structure the chitin chains are oriented in parallel fashion. The γ-structure is arranged in sets of three chains: two parallel and one anti-parallel. α-Chitin occurs in arthropod cuticles and is the most common arrangement; β- and γ-chitins occur in cocoons. α-Chitin is the most stable form and β- and γ-chitins can change to the α-form after various treatments.

Recently a new model has been proposed for chitin-protein complexes based on an X-ray diffraction study of ichneumon fly *(Megarhyssa)* ovipositor. The data support a model of a supercoil of 15-30 chitin chains as the core of a microfibril that is surrounded by a protein sheath, giving a total fibril diameter of 7.25 nm. The protein is composed of subunits spaced around the chitin core in a 6, helix. The chitin-protein fibrils are arranged in a hexagonal network and exhibit a fiber repeat of *ca.* 15.3 nm, or a multiple thereof, possibly due to additional supercoiling. The fly ovipositor is thus seen as a highly oriented structure in which both chitin and protein components are ordered at the molecular level, making X-ray diffraction patterns discernible. The structure of chitin-protein complexes in cuticle may not be as regular as in this specialized organ, as cuticle proteins are known to differ depending on the type and age of the tissue. Cuticle structure may therefore prove to be too complex to be elucidated by the X-ray diffraction technique, and other physical techniques that have the capability of analyzing less ordered chitinous structures are needed. Other types of studies have corroborated the protein-wrapped-around-chitin structure. For example, it was found necessary to use the sequential action of protease followed by chitinase to disrupt the larval cuticle of corn earworm.

CHITIN SYNTHESIS

Enzymatic processes that control the synthesis and breakdown of chitin play important roles in the growth, differentiation, nutrition, and behavior of insects. Researchers are slowly coming to understand how nature manipulates chitin in some organisms, in particular yeasts and fungi. Other species, including insects, are less well understood.

Chitin synthesis takes place via a metabolic pathway of eight or more steps converting hemolymph trehalose into the cuticle carbohydrate component. The pathway was first outlined using cellfree extracts of the mold *Neurospora crassa*. Since then studies using insect tissues have elucidated the same general sequence, in addition to identifying a lipid intermediate, dolichyl diphosphate-N-acetylglucosamine, that may also be involved. For example, the epidermal cells of the desert locust *(Schistocerca gregaria)* contain all of the enzymes necessary to convert trehalose or glucose to uridine-diphosphoN-acetylglucosamine; similar results were obtained using cultured imaginal wing disks from *Drosophila*. A whole-animal preparation of the Southern armyworm *(Prodenia)* has been used to assess the activity of chitin synthetase (UDP-2-acetamido-2deoxy-D-glucose: chitin 4-fl-acetamidodeoxy-Dglucoyltransferase) at various developmental stages. Synthetase activity was maximal late in the last instar and during the first 24 h of the pupal stage. More recently, an assay system using freshly dissected cuticle tissue has been described. Which of these enzymes is rate-limiting for chitin synthesis is unknown.

Most of the detailed knowledge of the chitin synthetase enzyme has been obtained from studies of yeast and fungal systems. Chitin synthetase zymogen has been isolated from yeast in an almost completely inactive form. It is a very large, multisubunit protein that is activated by a protease (activating factor, AF) which appears to be identical to the proteinase B and tryptophan synthetase inactivase II identified in other yeast studies. AF appears to be a serine protease and is found in vesicles within the yeast cell. The action of AF can be completely blocked by the presence of an inhibitor (AFI) which has been purified. AFI is a heat-stable protein with an apparent molecular weight of 8.5×10^3 daltons. AFI apparently binds to and directly inhibits AF since AFT has no effect on preactivated chitin synthetase. Whether a similar activating mechanism occurs with insect chitin synthetase is unknown. Certain inhibitors may actually be hydrolytic enzymes. An "inhibitor" was isolated from fungal cytosol that turned out to be an endochitinase.

Yeast chitin synthetase zymogen is uniformly distributed on the plasma membrane. The primary septum that develops between the mother and daughter cells is composed almost exclusively of chitin and is synthesized by local activation of the synthetase zymogen, presumably by directing AF-containing vesicles to appropriate sites on the membrane. This may be a general mechanism for morphogenic processes requiring growth in a specific direction and could be an important consideration for insect cuticle. The epidermal cell secretes and assembles components of cuticle on only one of its surfaces, resulting in precise spatial regulation of cuticle chitin synthesis.

In crustacea the relative rates of incorporation of glucose and N-acetylglucosamine are similar to those in locusts; glucose uptake is maximal after ecdysis and uptake of N-acetylglucosamine is maximal

TABLE 4.1 : CHITIN SYNTHETASE IN UNICELLULAR ORGANISMS

Organism	*Comments*
Blastocladiella emersonii (watermold)	Sensitivity of first hexosamine pathway specific enzyme to end-product inhibition is *not* constant throughout life cycle. Vesicles derived from gamma particles resemble chitosomes and synthesized chitin when supplied with UDP-GlcNAc.
Candida albicans (dimorphic fungus)	Cytoplasmic component isolated that inhibited activation of chitin synthetase; MW = 400 K; activity increased by treatment with phosphatidyl serine.
	Chitin synthetase activity in two polyene-resistant, ergosterol-deficient strains higher than parent strain. Chitin synthetase zymogen from mutant also more susceptible to trypsin digestion.
Mucor racemosus (yeast)	Chitin and chitosan syntheses in yeast and mycelial morphology compared and found to be three-fold higher in mycelial cells; modification of newly synthesized chitin occurs in both forms.
Mucor rouxii (yeast)	Cell wall microfibrils synthesized by soluble chitin synthetase in absence of cell membrane.
	Chitin synthetase particles (chitosomes) and microfibrils studied with EM; regard chitosomes as cytoplasmic conveyors of synthetase to cell surface.
	Chitin synthetase purified 120-fold; chitosomes with MW > 7 × 10^6; required proteolytic activation. Amphotericin-B and nystatin tested as inhibitors of chitin synthetase; Am-B K_i = 0.13 mM, non-competitive.

	Inhibition study with nucleoside—peptide antibiotics. Polyoxin A K_i = 0.6 μM; Nikkomycin X K_i = 0.5 μM; Nikkomycin Z K_i = 3.5 pM, all competitive; Polyoxin A does *not* exhibit antifungal activity, nikkomycins do. Digitonin stimulates chitin synthetase at low concentrations and inhibits at high concentrations; also causes release of 16S chitosome subunits with chitin synthetase activity.
Neurospora crassa (fungus)	Tunicamycin inhibited chitin synthetase in competitive manner with K_i = 480 μM.
Phycomyces (grease mold)	Benzoylphenylurea insecticides may act on activation of chitin synthetase zymogen and not directly on synthetase enzyme.
Rhodotorula glutinis (yeast)	*pH* change from 4.5 to 2 stimulates chitin synthesis; no proportional value in between decreases chitin content or increases total hexosamines.
Saccharomyces (yeast)	Yeast primary septum development sketched from EM studies.
	Chitin synthetase zymogen distributed over yeast cell membrane; scar formed from local activation of zymogen.
Saccharomyces carlsbergensis (yeast)	Chitin synthetase pH_{opt} 6.2; K_m for UDP GlcNAc 0.6-0.9 mM, polyoxin A a potent inhibitor.
	Localization of chitin in bud scars studied by EM, before and after action of specific enzymes.
	Purification of protein inhibitor of chitin synthetase activating factor, MW - 8.5 K.
Saccharomyces cerevisiae (yeast)	Heat-stable protein inhibitor of activa- -ting factor characterized.
	Purification of protein inhibitor of chitin synthetase activating factor.

		Chitin synthetase activating factor purified with affinity chromatography; MW = 44 K; pH_{opt} = 6.5 – 7; serine protease inhibited by PMSF.
		Assay method for chitin synthetase suitable for kinetic experiments described; mechanism is random or ping-pong. Temperature-sensitive mutant able to grow but cannot form buds. Defective gene product appears to be involved in selection of budding site and formation of chitin.

TABLE 4.2 : CHITIN SYNTHETASE STUDIES IN INVERTEBRATES

Organism	***Tissue***	***Comments***
Artemia salina (brine shrimp)	whole larvae and Mgt+	Chitin synthetase pH_{OPt} 7, endogeneous acceptor required; inhibited by UDP, diflubenzuron; not inhibited by polyoxin D.
Bombyx mori (silkmoth)	whole larvae	Radioactive tracer study after injection of [^{14}C]glucose. Glycogen synthesized early in feeding period was degraded and used as energy source and precursor in chitin synthesis.
Galleria mellonela (greater wax moth)	gut and peritrophic membrane	Mg^{2+} required for synthetase activity; inhibited by polyoxin D and UDP.
	epidermis	*In vitro* incorporation of GlcNAc followed; chitin synthetase preceded by elevated ecdysone titer.
Hemigrapsus nudus (purple shore crab)	integument early postmolt,	Chitin synthetase activity increases in premolt and K_m = 1.7 × 10^{-4}M for UDP GlcNAc.
Leucophaea maderae (cockroach)	leg regenerates	Sensitive quantitative assay for inhibition of chitin synthetase

		using leg regenerates; ID_{50} = 6.11 × 10^{-10}M for diflubenzuron and 7.53 × 10^{7}M for polyoxin D.
Locusta migratoria (grasshopper)	whole animal & abdominal integument	Incorporation of ^{14}C precursors of chitin followed; prior to imaginal ecdysis chitin precursors originated from old cuticle.
Manduca sexta (tobacco hornworm)	fifth stadium larvae	Diubenzuron (ID_{50} = 1.1 x 10^{-9} M) inhibited glucose and glucosamine incorporation into chitin; decrease in cuticle production achieved when fed or applied topically.
	cuticular disks	A short-term culturing method is described for analysis of chitin biosynthesis in larval cuticle. Incorporation of [^{14}C]GIcNAc followed.
Melanoplus sanguinipes (grasshopper)	abdominal integument	*In vitro* assay system for chitin synthesis labeled glucose, GIcNAc or UDP-G1cNAc as substrate; linear rate for 2 h.
	integument	Results consistent with chitin synthesis pathway in fungus; chitin synthetase accessible to substrate only from cell exterior.
Orconectes obscurus (crayfish)	whole animal	Incorporation of [^{14}C] glucose into chitin studied; chitin synthesis occurs in all stages of molt cycle; rate of chitin synthesis increases in premolt and peaks after molt.
Orconectes sanborni (crayfish)	whole animal	[^{3}H]glucose and [^{14}C]GIcNAc incorporated into chitin.
	epidermis, midgut gland	Early enzymes in chitin synthesis pathway, and not late

	and muscle extracts	enzymes, increase in premolt causing accumulation of UDP-GlcNAc and GlcNAc-6-P; late enzymes increase in postmolt.
Pieris brassicae (cabbageworm)	fifth stadium larvae	Polyoxin D or diflubenzuron injected and cuticle examined; total inhibition of cuticle growth and impairment of chitin synthesis noted.
Plodia interpunctella (Indian meal moth)	whole larvae and pupal epidermis	*In vitro* incorporation of GlcNAc followed; peaks of chitin synthesis at beginning of last-instar prepupae and in white pupae; exposure to ecdysone stimulated chitin synthesis.
Stomoxys calcitrans (stable fly)	pupae	Incorporation of radiolabeled GlcNAc in pupae studied; peaks at I and 4 days post-pupation, coinciding with production of ecdysial membrane and imaginal cuticle, respectively.
	pupae	Chitin synthetase isolated from pupal homogenates; pH_{OPt} 6.5; divalent cations not required; K_m = 31.7 pM and V_m = 135 pmol h^{-1} mg^{-1} for UDP-GlcNAc. GlcNAc, UTP, UDP, glycerine and N,N'-diacetylchitobiose inhibit enzyme; no inhibition by diflubenzuron.
	pupal epidermis	Assay system for studying chitin synthesis described; suitable substrates are radiolabeled glucose, glucosamine, fructose and GlcNAc; no MH required.
	whole animal	Tunicamycin and five insect

		growth regulators tested on pupal chitin synthesis; no inhibition noted.
Tetranychus urticae (two-spotted spider mite)	oocytes and hypodermis	EM study; role of cellular structures in chitin synthesis; nikkomycin used as inhibitor. Two separate systems described, one in oocyte and other in hypodermal membrane.
Triatoma infestans (blood-sucking bug)	whole animal and crude homogenates	Maximum incorporation of precursors into chitin occurs just after ecdysis; N-acetylglucosaminyl-phospholipid involved in chitin synthesis; tunicamycin inhibited conversion of GlcNAc to chitin; dolichyl diphosphate N-acetylglucosamine suggested as a chitin precursor.
Tribolium castaneum (red flour beetle)	larvae	Chitin synthetase requires Mg^{2+}; inhibited by polyoxin D, UDP and nikkomycin.
	gut	Chitin synthetase not inhibited by diflubenzuron and SIR8514 in *in vitro* assay.
Tenebrio molitor (yellow mealworm)	gut and peritrophic membrane	Mg^{2+} required for chitin synthetase activity; inhibited by polyoxin D and UDP.

before ecdysis. Some researchers believe that in crustaceans glycogen also serves as a source of precursors for chitin synthesis. Study of chitin precursors in crayfish has suggested that early enzymes in the chitin synthetic pathway increase in activity during the premolt period, causing a build-up of *N*acetylglucosamine-6-phosphate and UDP-Nacetylglucosamine. The activity of later enzymes increases during the post-molt period, causing a rise in overall chitin synthesis.

Cell-free biosynthesis of crustacean chitin has been achieved using a microsomal preparation from the brine shrimp *Artemia salina*. The enzyme preparation was able to catalyze the transfer of UDP-N-acetylglucosamine to an endogenous acceptor. The product was identified as chitin since it was degraded by purified chitinase and resisted extraction with alkali and urea. The synthetase appears to be located in the microsomal membrane.

Less information is known about the enzymes catalyzing chitin synthesis in insects because few, if any, of the enzymes and their effectors have been purified. The chitin synthetase of insects, like that in yeasts, appears to be an integral part of the cell structure. Even slight damage to epidermal cells leads to a shut-down of chitin synthesis. Incorporation studies done with radiolabeled chitin precursors show that insect chitin synthesis can be regulated at several points. Before ecdysis, *N*acetylglucosamine is used for the synthesis of new chitin, presumably because this substrate is produced by chitinase(s) hydrolyzing old cuticle. After ecdysis, prefomed N-acetylglucosamine is not readily available; glucose uptake is increased and the control point appears to be the amination of fructose-6phosphate. This is in agreement with the finding of Fristrom, J. that glucosamine and N-acetylglucosamine are sequestered in two separate metabolic pools. A detailed characterization of purified enzymes in the synthetic pathway for insect chitin remains to be done.

CHITIN DEGRADATION

β-Glucosaminidases are among the most important carbohydrate-splitting enzymes. They catalyze the hydrolysis of β-glucosamine linkages and their substrates can be oligosaccharides, polysaccharides, glycoproteins, mucopolysaccharides or mucolipids. One type of glucosaminidase is the β-*N*acetylglucosaminidases. In many animals they are important lysosomal enzymes whose deficiency leads to cellular pathology and clinical disorders such as Tay-Sach's and Sandhoff s diseases. In insects β-*N*acetylglucosaminidases are also molting enzymes whose excess or deficiency may lead to developmental arrest. The substrates for the insect enzymes are chitin, the primary structural polysaccharide in cuticle and peritrophic membrane, and chitin oligosaccharides. The catabolism of chitin is catalyzed by two specialized β-glucosaminidases, endochitinase [poly(1,4-β-(2-acetamido-2-deoxyD-glucoside)) glycanohydrolase] and exochitinase [oligo(1,4β-(2-acetamido-2-deoxy-D-glucoside) glycanohydrolase].

These specific citations are not meant to be complete and all-inclusive. In many of these reports it is hard to ascertain if hexosaminidase or true chitinase activity was being studied, often due to the conditions chosen for the enzyme assay. A substrate (artificial) may have been used which masks the true specificity such that the *in vivo* function was not clearly defined. Hexosaminidase (chitinase) was found in many animals in different organs/ locations, including integument, molting fluid, hemolymph, saliva, digestive juice, midgut, silk gland, fat body and testis; often it was present in separate organs within the same animal. In a few cases activity was clearly shown to belong to multiple enzymes. Many of the enzymes parallel the larval-pupal-adult transformation and therefore probably play some role in metamorphosis. Others do not fluctuate as such, and function in different capacities such as digestion. Midgut chitinase may hydrolyze food sources containing chitin; it may also have lysozomal enzyme functions such as glycoprotein and glycolipid processing.

Molting Fluid

Molting fluid digests the old endocuticle, thus weakening the rest of the cuticle in preparation for its mechanical rupture at the time of ecdysis. After detachment and retraction of the epidermis from the overlying cuticle (apolysis), the molting space formed in some insects is filled with a colorless and viscous secretion, the molting gel. Soon thereafter the epidermis begins to secrete a new cuticle which thickens as a result of sequential deposition of layers, and hardens as a result of sclerotization. Also, a marked decrease in the viscosity of molting gel occurs, forming the molting fluid. At this time, digestion of the old cuticle begins and, within a time period of several hours to a few days, the endocuticle is completely degraded. The molting fluid is then gradually resorbed and only the thin sclerotized outer layers of old cuticle remain. These are mechanically ruptured at ecdysis and discarded as the exuviae.

Many of the organic and inorganic components of molting fluid in two silkmoths and the tobacco hornworm have been determined. Larval-pupal molting fluid of *Hyalophora cecropia is* a hyperosmotic salt solution relative to hemolymph with potassium representing the principal cationic and osmotic component. Bicarbonate is the principal anionic component. Pupal-adult molting fluid from *H. cecropia is* similar except that it is hypo-osmotic relative to hemolymph while larval-pupal fluid from *Manduca sexta* is iso- or slightly hypo-osmotic. The pH of the molting fluids characterized varies from 7.2 to 8.0 and appears to depend on the method of anesthesia used to quiet the animals. Carbon dioxide anesthesia results in a more alkaline pH (pH 7.7 versus pH 7.2 for molting fluid from non-anesthetized animals). Molting fluid contains proteins, polypeptides and amino acids but, interestingly, no sugars.

The idea that molting fluid contained enzymes capable of digesting cuticle was postulated early in this century. In view of the chitin-protein nature of endocuticle, it is not surprising that both chitinases and proteinases are present. Two proteinases are present in fluid from *Antheraea polyphemus* pharate adults. A single exochitinase and several endochitinases occur in molting fluid from *Manduca sexta* pharate pupae. Esterases are also present, including carboxylesterases and arylesterases. Two esterase fractions were separated from the molting fluid of *A. polyphemus* while three were identified in *M. sexta*. Molting gel lacks proteolytic activity but does contain an inactive form of proteinase (proenzyme) that can be activated. Analogously, it may contain an inactive form of chitinase (proenzyme?) that becomes activated in the molting fluid. Tyrosinase may also be present.

Chitinase Assays

There are many substrates available for assaying chitinolytic enzymes. The ideal substrate, of course, is chitin itself. However, chitin is insoluble and preparations frequently contain undefined chemical constituents such as amino acids, peptides or proteins. Thus, it is very difficult to obtain a substrate that truly reflects the native structure of chitin in cuticle and at the same time provides enough sensitivity so that low levels of enzymes are not overlooked.

Cuticle is a highly organized chitin-protein-lipid supramolecular structure. Studies on enzymes acting on artificial or modified substrates are difficult to interpret in physiological terms because the kinetic results *in vitro* may demonstrate only a part of the enzyme's true specificity and catalytic efficiency. Modified chitin and pseudochitin substrates suffer from such shortcomings, although some pseudosubstrates offer the advantages of increased sensitivity and ease of assay. Many assay methods have been devised for detecting chitinase activity, none of which have the hoped-for combination of ease, accuracy and nativeness of cuticle chitin structure. Several of these procedures will be described.

Endochitinase is measured primarily on large polymeric substrates. Colloidal chitin has been prepared from crustaceans, insects and fungi and it can be utilized as a chitinase substrate in several ways. The decrease in turbidity of the substrate can be used as a direct measurement of enzymatic activity or the amount of Nacetylglucosamine released can be measured colorimetrically after enzyme inactivation. The latter result is not reliable because it actually measures the activity of exochitinase whose substrates (small oligosaccharides) are the products of endochitinase catalysis. Chromogenic assay of the N-acetylglucosamine released from deproteinized cuticle squares has been used to assess chitinase activity. Although these assays have the advantage of more "nativeness of substrate", they are insoluble substrates whose chemical structure is very poorly defined. The catalysis then, is a heterogeneous one, and such experiments are unable to characterize precisely the enzyme's specificity because controlled kinetic studies are practically impossible. Some chitinase enzyme sources are devoid of exochitinase such that no N-acetylglucosamine release occurs. To circumvent this potential problem, C. Jeuniaux's method of assay includes the addition of exochitinase either as lobster serum or as a reconstituted pure enzyme.

Glycol chitin, a commercially available modified chitin that is soluble in aqueous solutions at physiological pH, has been used as a substrate for chitinase. Its activity can be detected by repeatedly testing the viscosity of the substrate/enzyme mixture. This method can be used to distinguish exoand endo-chitinase activities as the endoenzyme will cause a much more rapid decrease in the viscosity of the substrate solution. However, performing the viscometric assay is tedious. The use of potassium ferriferrocyanide as an oxidizing agent to measure the generation of reducing end groups colorimetrically has made the glycol chitin assay for endochitinase easier to carry out. The degree of glycolation of chitin determines the solubility and susceptibility to enzymatic attack. Glycolation is done using ethylene oxide or 2-chloroethanol, the latter being preferred because of the difficulty of preparing the former reagent. Deacetylated glycol chitin is produced under the conditions of chitin glycolation, and it is necessary to reacetylate the product with acetic anhydride.

Two chitinase assays take advantage of the solubility of N-acetylglucosamine oligomer/ monomer products which are prepared from the insoluble chitin. One uses tritiated chitin, either prepared *in vivo* or by reacetylating chitosan with tritiated acetic anhydride. Cleavage of the labeled substrate is detected as radioactivity soluble in 10% trichloroacetic acid. The second assay uses Chitin Red® (Calbiochem) as substrate, a preparation of γ-chitin coupled to a red chromophore. Chitinase activity produces a redcolored supernatant.

Chitin oligomers have been used as substrates in kinetic studies of non-specific chitinases such as lysozyme to measure both endo- and exochitinase activities. The chitin oligomers ranging in size from disaccharide to heptasaccharide are prepared by hydrolyzing crab or shrimp shells with concentrated hydrochloric acid and separating the oligomers with either charcoalcelite or gel permeation chromatography. As a precaution, chitin oligosaccharides should be treated with an acylating agent such as acetic anhydride to ensure that no chitosan oligomers are generated. Endo- or exo-cleavage patterns can be determined by following the reaction time-course. N-Acetylglucosamine will be an initial product only if the exoenzyme is present. The reaction products are detected by these same chromatographic methods, paper chromatography, thin-layer chromatography or highperformance liquid chromatography. Radiolabeled oligosaccharides or oligomers oxidized at the non-reducing end by periodate treatment may also be used. If a need for glucosamine or chitosan oligomer analysis arises, they can be identified and quantified using a conventional amino acid analyzer and either ninhydrin, o-phthaldialdehyde, or fluorescamine for derivatization.

An unusual chromophoric oligosaccharide substrate, 3,4-dinitrophenyl-tetra-N-acetyl-β-chitotetraoside, distinguishes endo- from exoenzyme activity. Exochitinase will release dinitrophenol either immediately or following a lag phase after incubation with substrate, depending on whether hydrolysis proceeds from the reducing or non-reducing end of the oligosaccharide. Endochitinase does not release dinitrophenol, but it will generate smaller dinitrophenylated sugar products.

Exochitinase activity can be assayed using chitin pseudosubstrates containing N-acetylglucosamine and a chromophore (e.g. p-nitrophenol) or fluorescent component (e.g. 4-methylumbelliferone). The increase in absorbance or fluorescence can be followed continuously at a discrete wavelength through the course of the reaction. These substrates are commercially available and the assays are sensitive, quick and relatively inexpensive. Structurally related compounds are also available that can be used to study in detail structure-activity relationships and the mechanism of catalysis.

Exochitinases β [exo-b(1 → 4)-oligo-Nacetylglucosaminidases] and Endochitinases [endo-β (1 → 4)-poly-N-acetylglucosaminidases]

Invertebrate enzymes that hydrolyze substrates which contain N-acetylglucosamine as part of their structure have been studied for many years. Although the importance of chitinolytic enzymes in the growth and development of arthropods has been recognized for many years, little work has been published on the properties of purified enzymes. Most of the detailed knowledge of the chitinolytic enzymes has come from studies of fungal and yeast enzymes. In many cases concerning insects there is ambiguity about the kind of activity measured (exo- or endo-) and the homogeneity of the enzymes. Primarily, heterogeneous enzyme preparations and/or poorly defined substrates were used. Several of the more definitive studies of chitinolytic enzymes will be described here, including those on the silkworm, *Bombyx mori;* the tobacco hornworm, *Manduca sexta;* the fruitfly, *Drosophila hydei;* the migratory locust, *Locusta migratoria* and the hunting spider, *Cupiennius salei*.

Forty years ago chitinase activity was noted in cast exuviae and in molting fluid of *Bombyx mori*. Subsequently, Kimura, S. has conducted very elegant studies of the chitinolytic enzymes in silkworm tissues during larval and pupal development. Endoand exo-β-N-acetylglucosaminidases were detected in integument with the exo-activity increasing several-fold after injections of 20-hydroxyecdysone. Molting fluid contained three exo- and two endoenzymes that were separated by chromatographic methods. Similar enzymes were found in midgut, fat body, testis and silk gland. Larval hemolymph contained an antigenically distinct exo β-N-acetylglucosaminidase and no endo-enzyme. Molting fluid and integument contained endochitinase and chitobiase, the latter cleaving only $\beta GlcNAc_z$ and not chitin. Larger oligosaccharides were not tested.

Another insect whose chitinolytic enzymes has been studied extensively is *M. sexta*. Bade, M. and Stinson, A. have proposed that hornworm chitinase hydrolyzes cuticle chitin after proteases in molting fluid unmask the chitin from bound proteins. Smith, *R. et al.* report that digestion of chitin-protein ghosts derived from 1% sodium dodeyl sulfate-boiled extracts of *Heliothis zea* larvae occurred using a combination of a proteolytic enzyme followed by chitinase whosesource was not revealed. Conversely, Lipke, H. and Geoghegan, T. found that native chitin in *Sarcophaga bullata* was not made more available to chitinase by prior treatment with proteases. Bade and her colleagues also suggested that molting fluid contains a single chitinase which exhibits positive cooperativity and acts via a processive mechanism, cleaving off successive GlcNAc monomers from the same chitin chain without diffusing from one polymer to another. In all of these studies, crude molting fluid and insoluble substrates were used. Since multiple chitinolytic enzymes are known to be present in molting fluid, it is difficult to interpret the kinetic data derived from such a heterogeneous catalysis. These conclusions must be regarded with caution until studies with purified enzymes and better substrates are conducted.

Kramer and colleagues have used purified enzymes and well-defined soluble substrates to determine the kinetic behavior of the chitinolytic enzymes from the tobacco hornworm. Their studies, and those of other laboratories, suggest that two classes of enzymes, exochitinase and endochitinase, degrade cuticle chitin in insects. One exo-β-N-acetylglucosaminidase (ExI) occurred in molting fluid and pharate pupal integument and a second exoenzyme (ExII) was present in larval hemolymph. Both enzymes were found in pupal hemolymph. Either enzyme could hydrolyze oligosaccharide substrates following an exo-cleavage pattern beginning at the non-reducing end. Their physical, chemical and kinetic properties were very similar with one major exception. ExI was more efficient at hydrolyzing βGlcNAc oligosaccharides. ExII was at least two orders of magnitude less active. The two enzymes were unrelated immunologically. The tissue localizations, together with the kinetic behavior, demonstrated that the former is a true exochitinase and that the latter is an exo-β-Nacetylhexosaminidase with a "sluggish" exochitinase activity. ExI degraded substrates as large as hexamer (and probably larger ones) to monomer which can be recycled *in vivo* and utilized as a precursor for the chitin formed in cuticle of the next developmental stage.

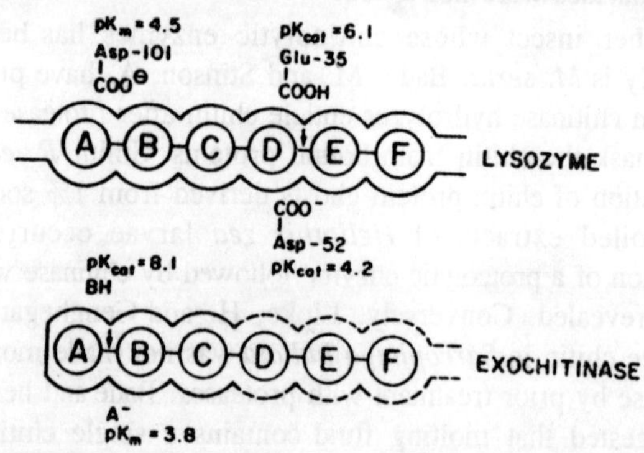

Fig. 4.4. A comparison of the proposed active site of M. sexta exochitinase with that of hen's eggwhite lysozyme. β (1-4) GlcNAc hexasaccharide denoted by A - F with A being the non-reducing end.

pH rate studies of *M. sexta* exochitinase suggested that two ionizable groups in the active site with apparent pK_a values of 3.8 and 8.1 are important for binding and hydrolysis, respectively. The identities of these residues are unknown, but speculations can be made based upon an examination of another β-N-acetylglucosaminidase, hen's eggwhite lysozyme. Lysozyme has an acidic group, aspartic acid 101 (pK_a = 4.5), involved in a binding contact with substrate in subsite A that must be ionized to interact. Exochitinase may have a similar acidic group with a slightly lower pKin its active site. Two other carboxylic acid side-chains participate in acetal bond hydrolysis in lysozyme (glutamic acid 35 and aspartic acid 52), but apparently only one other group in exochitinase. There appears to be no counterpart in exochitinase to lysozyme's aspartic acid 52, which stabilizes a carbonium ion or covalently displaces the leaving group in the lysozyme mechanism. There is a protonated group with a relatively high pK in exochitinase that may participate as a proton donor analogous to the role played by glutamic acid 35 in lysozyme. Whereas lysozyme exhibited a rather narrow pH optimum (pH 4-6), *M. sexta* exochitinase had a pH optimum extending into the alkaline range (pH 4-8) which is due to the basic group, BH, in the active site. Candidate residues for this basic group include histidine and lysine. A similar pH-activity profile was found for another exo-β-N-acetylhexosaminidase from a fungus, *Aspergillus niger.*

X-ray crystallographic studies have shown that the active site of lysozyme is an open cleft with six subsites. The binding site of *Manduca sexta* exochitinase may also have subsites, probably only three or four forming a close-ended cleft since trimer through hexamer had similar K_m values. Exochitinase exhibited substrate and product inhibition effects *in vitro,* indicating that a large active site exists for both productive and nonproductive modes of binding. Whether these inhibition phenomena happen *in vivo is* unknown.

The substrates for exochitinase are generated from cuticle chitin by endo-β-acetylglucosaminidase or endochitinase. Three endo-β-Nacetylglucosaminidases were purified from the integument of *M. sexta* pharate pupae. These enzymes are glycoproteins that have unique molecular weights, 7.5 × 10^4 for E_NI, 6.2 × 10^4 for E_NII and 5.0 × 10^4 for E_NIII, and also contain neutral hexoses and hexosamine. Molting fluid contained mostly ENI, and smaller amounts of E_NII and ENIII. The latter was immunologically related to E_NII but not to E_NI. Analysis of initial velocity experiments and products formed from β (1 → 4) GlcNAc oligosaccaride and glycol chitin substrates revealed that the longer substrates were preferred and that all three enzymes behaved primarily as endochitinases. Oligosaccharides rapidly disappeared until only disaccharide and trisaccharide remained, after which a small amount of free N-acetylglucosamine was produced from trisaccharide very slowly. Thus, the enzymes exhibited a "sluggish" exochitinase activity, but only with chitotriose as substrate. No activity was expressed toward pNpβGlcNAc, $βGlcNAc_2$ or *Micrococcus lysodeikticus* cell walls. The endochitinases were also susceptible to substrate inhibition. The active site of these enzymes is probably a large open cleft, much like that of lysozyme, which accommodates both productive and nonproductive binding of substrate.

Endochitinolytic enzymes from various sources differ widely in their physical and kinetic properties. Insect endochitinases occur in several tissues and sizes. Mammalian sera contain enzymes that are about the same size, 5-6 × 10^4, while plant endochitinases are smaller, 3 × 10^4. Whether the mammalian and plant enzymes are true chitinases *in vivo* is uncertain. The pH optimum appears to depend on the enzyme source and substrate used. Low optima (-pH 2) were observed with yeast, goat and bovine endochitinases when glycol chitin or radiolabeled chitin was the substrate. However, the latter two enzymes showed maximum activity with colloidal chitin at the same pH as that found for the *M. sexta* enzymes (pH 6). Fungal and plant endochitinases also had pH optima of 6. Bade, M. and Stinson, A. reported that chitinolytic activity in *M. sexta* molting fluid was maximum at a slightly higher pH, 7, but that it was more reproducible at pH 6.5. Raw molting fluid and insoluble chitin were used in their experiments, however. Similar K_m values (mg ml^{-1}) are exhibited for glycol chitin by mammalian and insect enzymes.

A proposed mechanism ofchitin catabolism by the two types ofchitinolytic enzymes in *M. sexta*. After endochitinase (EN) hydrolyzes chitin (βGlcNA$c_{n>}$,) to yield oligosaccharides (βGlcNAc_n), the latter may bind to exochitinase (E_x) to give a productive complex (E_x./ iGlcNAc_n)$_p$ where the substrate is susceptible to the action of the catalytic site. At higher concentrations additional substrate molecules may also bind either enzyme such that none of the bonds are properly exposed to the catalytic amino acid functional groups and nonproductive complexes are formed ([E,.(ri + 1) β G1cNAcn]np). Compounds may inhibit E_x in a competitive pattern where more than one inhibitor may bind (Ex-In) depending on the inhibitor's structure and concentration. The oligosaccharides may be sequentially degraded by exochitinase from the non-reducing end to yield monosaccharide (n β G1cNAc). This kind of mechanism is not unusual for enzymes that modify polymeric and oligomeric substrates, and models for such hydrolyses have been proposed that envisage substrate interacting with the entire active site (productive complex) or only parts of it (non-productive complex).

Chitinolytic enzymes have also been partially characterized from integument of *Drosophila hydei* and *Locusta migratoria*. Endochitinase and exo-β-N-acetylhexosaminidase from *D. hydei* exhibited pH optimum between 5.5 and 6.2 and temperature optima of 35° and 50°, respectively. Three endoenzymes and one exoenzyme fraction were partially purified by gel filtration, which is identical to the number of enzymes found in *M. sexta* integument. Exo activity was also detected in *Drosophila* gut and fat body. The pH optimum for *Locusta* integument endochitinase and exo-β-N-acetylhexosaminidase was between 4 and 5. Temperature optima were 50° and 70°, respectively. Both endo- and exo-activities reached maximum levels during the latter part of the molting cycle and declined before ecdysis.

Some of the most active hydrolytic enzymes in the digestive fluid are chitinolytic ones. These probably digest chitin and other carbohydratecontaining food sources. Two of the bettercharacterized digestive chitinases were obtained from the digestive juice of the hunting spider, *Cupiennius salei*. These are similar to the integument enzymes found in insects.

An endochitinase ($MW_{app} = 4.8 \times 10^4$) degraded chitin to disaccharide and trisaccharide. Exochitinase ($MW_{app} = 1 \times 10^5$) hydrolyzed the oligosaccharides to monomer.

The reason for multiple chitinolytic enzymes in insects is unknown. The enzymes may act in a concerted mechanism or individually. In mammalian fibroblasts, N-acetylhexosaminidases are synthesized in the form of precursors of larger molecular weight and are shortened later to the size of chains found in tissues. This processing and additional posttranslational modification (phosphorylation) allow recognition for incorporation into lysosomes. Whether these modifications occur with insect enzymes is unknown.

Lysozymes

In addition to its bacteriolytic action, lysozyme (mucopeptide N-acetylmuramolylhydrolase) also cleaves chitin substrates. This enzyme may be mistaken for a chitinase and it is imperative to assay suspected enzymes for lytic activity (muramidase) towards bacterial cell walls such as *Micrococcus lysodeikticus*. Chitinase does not exhibit such activity.

Several insect lysozymes, as well as some other invertebrate ones, have been characterized. They are generally small basic proteins (approx. M.W. 1.5-2 $\times 10^4$) with properties (heat stability, pH and ionic strength optima) similar to those of vertebrate lysozymes. Tissue sources include eggs, gut and hemolymph. Insect lysozymes are probably antibacterial in function because enzymatic activity increases dramatically when the animal is challenged with bacteria.

Insect Cell Lines

Several insect cells exhibit chitinolytic enzyme activity. The first reported were two cell lines from *Periplaneta americana* whose spent media contained colloidal chitin degrading activity and also hydrolyzed bacterial cell walls. It was subsequently shown by chromatographic methods that lytic and chitinolytic activities were distinct. β-N-Acetylhexosaminidase activity was detected in culture media from cells derived from *Culex quinquefaciatus, Drosophila melanogaster* and *Culex fumiterana,* but not from *Trichoplusia ni* or *Aedes pseudoscutelaris*. The *Culex* medium hydrolyzed βGlcNAc oligosaccharides in a manner that demonstrated the presence of exo- and endochitinases.

REGULATION OF CHITIN METABOLISM

The complex structure of insect cuticle, and the difficulty in obtaining stable, purified preparations of insect chitinase and chitin synthetase, have complicated the study of control mechanisms in chitin metabolism. Because chitin is rarely found without surrounding protein, the study of chitin synthesis and degradation is of necessity linked to the study of the whole cuticular structure. Some of the papers cited here will, therefore, discuss events in whole cuticle as a way of observing changes in chitin suprastructure. The recent developments in genetic studies will soon make the investigation of individual cuticle components technically feasible. Studies concerning chitin metabolism divide themselves into two categories: those investigating hormonal and developmental modes of control and those investigating other than hormonal regulators, such as pesticides and insect growth regulators. These modes of control will be considered separately.

Each insect molt is a response to the hormonal and enzymatic milieu present at the time previous to and concurrent with the event. Hormonal control of insect growth and development has been reviewed extensively. A variety of hormones, many of them peptides and still uncharacterized, control the homeostatic events in the life of an insect. Research of many years has shown, though, that the cyclical process of growth and molting is controlled primarily by discrete quantities of three or four hormones which are secreted at critical time points in development. They are prothoracicotrophic hormone (PTTH), molting hormone (MH), juvenile hormone (JH), and eclosion hormone (EH). These are discussed in detail in other chapters in this series.

The effects of MH and JH on chitin biosynthesis have been studied in various species using metabolite incorporation experiments. In both *Triatoma infestans* and *Locusta migratoria* maximum incorporation of radiolabeled precursors into chitin occurred after ecdysis. Incorporation peaks have also been noted post-pupation. Administration of 20hydroxyecdysone causes a decrease in the rate of [^{14}C]glucose incorporation into chitin in *Bombyx mori,* followed by increased incorporation until ecdysis. Exposure of *Plodia interpunctella* larval epidermis to 20hydroxyecdysone stimulated chitin synthesis, but only after a 24 h delay following exposure to the hormone; this is consistent with the effects of MH observed in crayfish. Apparently a preincubation period is required before cells become competent to respond to hormone. Studies have also been done in which the change in integument enzyme activity patterns were correlated with the molt cycle.

The desire to study the developmental events of epidermal cells in a system in which exogenous effects could be better controlled led investigators to try tissue explant techniques. Successes were few, however, until the knowledge of insect biochemistry had increased. Gradually it was learned that a source of molting hormone was necessary to induce synthesis of cuticular components and cuticle formation and that 20-hydroxyecdysone was the active form of molting hormone. Studies have investigated the effect of hormone concentration, duration and interruption of dose. Marks, E., using cultured cockroach leg explants, was able to show that cuticle deposition could be induced up to five times in the same explant when the molting hormone dose was repeated at discrete intervals. Other studies looked at the role of JH and found that in *Manduca sexta* formation of larval cuticle depended on exposure to JH, at the proper titer, along with 20hydroxyecdysone. If JH was' absent, pupal cuticle was synthesized. The timing of cuticular events in cockroach leg regenerates has been studied by timelapse photomicrography. It has been noted that the deposition of cuticle in the Indian meal moth lasts only 2 days, but that the epidermal cells have a 4-day preparation period before this, during which protein is synthesized. Data concerning the control of chitin biosynthesis are consistent with the hypothesis that ecdysteroids promote the synthesis of RNA and protein followed by a period of ecdysteroid independent protein and chitin syntheses.

The authenticity of cuticle synthesized *in vitro* has been tested in a number of ways. The product varies considerably from system to system and photography from various experiments is displayed in Marks, E. and Sowa, B.. Chitin formed *in vitro* was tested for histochemically, with chitinase conjugated fluorescent stain, and was found in cultured cockroach leg regenerates. In another study radiolabeled N-acetylglucosamine was taken up by 20hydroxyecdysone stimulated leg regenerates. A synthesis and digestion study done on cultured wing discs showed that radiolabeled D-glucosamine was taken up into the epidermal and cuticular regions; chitinase treatment resulted in release of labeled material into the medium.

Other radiolabeled incorporation studies have shown that wing tissue from *Antheraea polyphemus* (silkmoth) can be induced by MH to specifically synthesize epidermal cell proteins. Cockroach embryos cultured *in vitro* with radiolabeled precursors and MH show that the localization of the synthesized cuticular products is a function of the stage of development of the epidermal cells. The many studies done on invertebrate tissue culture have been reviewed elsewhere.

INSECT GROWTH REGULATORS

Chitin is an ideal target for pesticide development because of its specialized occurrence in nature and its critical role at each stage of morphogenesis. An understanding of chitin enzymology may help to develop a new class of insect growth regulators that interfere with chitin metabolism. Either chitin excess or deficiency during any morphogenetic cycle can produce deleterious and lethal effects.

Natural enzyme inhibitors, especially inhibitors of hydrolases, are found in many plants, animals, bacteria and fungi. Some of these have been useful in the treatment of carbohydrate-dependent metabolic disorders such as diabetes mellitus. Hydrolases, including chitinase, found in plants and microorganisms, may act as defense mechanisms. Natural products that are used as fungal antibiotics have been described that interfere with chitin deposition. The polyoxins, structural analogs of UDPN-acetylglucosamine, have been studied in detail and their mode of action defined as a competitive inhibitor of chitin synthetase.

Although some fungicides and antibiotics have insecticidal activity in the laboratory, they have not been used as agricultural chemicals. However, since 1970 a group of insecticides has been developed that specifically interferes with chitin metabolism. The best-studied of these is Diflubenzuron® TH6040, [N-(4-chloropheny)-N'-(2,6-difluorobenzoyl)-urea], a substituted urea compound that apparently disrupts production of chitin in newly formed cuticle so that the insect has difficulty in molting and consequently dies. Histological studies note a variety of structural changes after the administration of benzoylureas, including an inability to incorporate [^{3}H]glucose into endocuticle, and an absence of normal endocuticle deposition. Diflubenzuron is commercially available and is currently used to control certain forest insects.

The exact mode of action of this and other antichitin compounds is unknown although it has been shown that both chitin synthesis and degradation are affected in insects. A structure/function study done on a group of 24 benzoylureas in the milkweed bug *(Oncopeltus fasciatus)* showed good correlation between toxicity and inhibition of chitin synthesis without alteration in the metabolism of ecdysone and 20-hydroxyecdysone. Because inhibition of chitin synthesis occurs first, this phenomenon rather than chitin degradation appears to be the predominant mechanism leading to insect death.

Unlike the polyoxins, the benzoylphenylureas may not act directly on chitin synthetase. Recent experiments show that these compounds are unable to inhibit chitin synthetase in a cell-free system but they do act effectively in the cockroach leg regenerate assay and a partially purified plasma membrane preparation. It has been proposed that these compounds act as serine protease inhibitors, preventing the conversion of chitin synthetase zymogen to active enzyme. However, no direct effect on a protease has been demonstrated. In nearly every inhibition experiment it has not been determined if chitin is the only carbohydrate affected. Levels of other metabolites, such as glycoproteins and mucolipids may also be decreased.

To date no cytotoxic effects of TH6040 have been noted in higher animals, nor has it been demonstrated to affect the synthesis of complex carbohydrates. However, toxic effects at 10-30 parts per billion have been noted in brine shrimp and some resistant strains of housefly have already developed.

Because of the critical protective function of cuticle, chitinases have not been overlooked as insecticides and antimicrobial agents, although no insecticide is known presently that specifically interferes with chitin degradation. A preparation of R-1,3 glucanase and chitinase has been tested and found effective as a treatment for rice blight. The insecticidal effect of *Bacillus thuringiensis* spores and toxin against the spruce budworm is increased when applied with the addition of chitinase. The combined preparations show an improved mortality rate and better foliage protection; the chitinase apparently facilitates entry of the pathogen by hydrolyzing the chitinous layer of the gut. Various preparations of *B. thuringiensis* and chitinase have been field-tested. Chitinase itself has been found to be both toxic and non-toxic to insects.

A number of chemicals have been tested for inhibition of chitin synthesis and degradation. Suffice it to say that this number is not very large, especially in regard to chitinase. Pesticide chemists have put forth relatively little effort to develop compounds directed against chitin metabolism. We have only begun to understand the enzymatic mechanisms involved in the synthesis and degradation of chitin. Hopefully, these results will stimulate greater effort to synthesize and test compounds that may perturb chitin biochemistry.

LIPIDS

The cuticular lipids of insects consist of aliphatic material which is present on the outer layer of the integument. These waxes serve the critical function of restricting water loss to prevent desiccation in most terrestrial insects, and also serve as sex attractants and aphrodisiacs, as species and caste recognition cues, and as kairomones in certain insect groups. The cuticular lipids affect the absorption of insecticides and other chemicals from the environment and have been suggested to serve as a barrier to the penetration of micro-organisms. Some insects, such as the scale insects and honeybees, produce much larger amounts of waxy material for protection and for raising brood and storing honey.

All insects which have been carefully studied possess hydrocarbons in their cuticular lipids, and they are the most abundant class of compounds in many species. Other types of waxes commonly found in cuticular extracts include wax esters, ketones, epoxides, secondary alcohols, primary alcohols, free fatty acids, sterols, sterol esters and triacylglycerols. In this context, wax is used in its broad sense to include a variety of lipids rather than in the strict chemical definition.

Cuticular lipids are synthesized by cells which are located near the integument or in the peripheral fat body, probably oenocytes. The biosynthetic pathways for the major cuticular components, including hydrocarbons, wax esters and secondary alcohols, have been established. Prior to deposition, the lipids must pass through the epidermal cells and then through the cuticle to reach the surface of the insect. The pore canals have been implicated in the process by which cuticular lipids cross the cuticle. The thickness of the cuticular lipid layer appears to vary considerably among different species, and values from a few molecules thick to over half a micrometer have been reported.

Various aspects of insect cuticular lipids have recently been reviewed. Discussions focusing on aspects of the chemistry and biochemistry of cuticular lipids have been presented by Blomquist, G. and Jackson, L; Nelson, D.; Lockey, K.; Jackson, L. and Blomquist, G.; Blomquist, G. and de Renobales, M. and Jackson, L. and Baker, G.. Analytical techniques used to characterize insect waxes have been reviewed by Jackson, L. and Armold, M. and Gilby, A.. The role of cuticular hydrocarbons in chemical communication has been dealt with by Howard, R. and Blomquist, G., and discussions of the role of cuticular lipids in limiting water loss have recently been presented by Hadley, N.; Gilby, A.; Machin, J.; Edney, E. and Ebeling, W.. This chapter summarizes the functions attributed to insect cuticular lipids, outlines analytical procedures used in characterizing wax components, reviews the types of chemicals commonly found in cuticular lipids, discusses the biosynthesis of cuticular lipids and reviews the physiological and endocrine regulation of cuticular lipid formation.

FUNCTIONS OF CUTICULAR LIPIDS

Prevent Desiccation

Perhaps the most critical of the functions attributed to insect cuticular lipids is the formation of a waterproof layer which prevents a lethal rate of desiccation. Many investigators have measured the transpiration of water from insects, and these studies have been reviewed periodically. The evidence is in agreement that the cuticle is permeable to water vapors, and that the cuticular lipids play a major role in reducing transpiration.

Major lines of evidence which support this view are:

(1) Cuticles from which the surface lipids are removed with organic solvents are relatively permeable to water. Vegetable oils, lecithin, and a series of wetting agents and detergents affect the permeability of the cuticle to water.

(2) The waterproofing observed with intact insects is closely duplicated when extracted cuticular lipids are deposited on collodium membranes or intact wing membranes. Rapid desiccation occurs as a result of scratching the outer surface of the cuticle with abrasive

dusts, and this has been used in insect control. Adsorption of the lipids onto the dust may also play a role in increasing transpiration.

(4) If the abraded insect is kept in a moist atmosphere to prevent desiccation, the lipid layer is restored, and along with it, the resistance to desiccation returns. Positive correlations have been shown between the quantity of extracted cuticular wax and the measured or potential transpiration. In addition, a greater amount of cuticular lipid is often present when there is a greater need for water conservation.

(6) Insects which inhabit desiccating environments have a surface-wax composition that provides maximum impermeability to water vapor.

(7) The transpiration rate from an insect is found to increase rather abruptly at a temperature that corresponds to the transition point or change of phase point of the lipids on the cuticle of the particular species.

This latter point has become quite controversial. The discovery that the permeability of insect cuticle underwent abrupt increases at specific temperatures prompted Beament, J. and Wigglesworth, V. to propose that cuticular permeability is controlled by a monolayer of polar lipids oriented at an angle of 65° to the cuticle surface. It was proposed that at the transition temperature (Tc), sufficient thermal energy is available to disrupt this ordered state, leading to increased transpiration. Another model proposed to account for the permeability changes at Tc includes M. Locke's suggestion that the molecules in the wax canals undergo a phase change at Tc.

Machin, J. (1980) and Hadley, N. have summarized the recent evidence which argues against the oriented monolayer model proposed by Wigglesworth and Beament. The major lines of evidence against the model are:

(1) The cuticular lipids of most insects studied to date are not composed primarily of aliphatic alcohols and acids, but rather contain large amounts of hydrocarbon;

(2) Forced area curve data for cockroach cuticular lipid indicated that a monolayer was absent and that the molecules were only weakly attracted to the water surface; Toolson, *E. et al.*, using spin-labeled molecules, showed that there was not a preferred orientation on the cuticle, and furthermore showed that the phase transitions take place with lipids associated with the surface rather than the pore canals;

(4) Toolson, E., using mathematical arguments, has shown that the abrupt permeability increases which occur at Tc when water loss is plotted against temperature are in reality artifacts created by

incorrectly dividing the ob served rate of water loss by the saturation deficit to correct for thermally induced changes in the diffusion gradient.

Using Toolson's method, Machin, J. calculated data for a number of insects, and suggests that the presence or absence of temperature transitions depends upon whether the insect has a "greasy" or "waxy" cuticle. A generally accepted model for the waterproofing observed by cuticular lipids awaits development.

Effect on Entry of Insecticides and Micro-organisms

The cuticular lipids play a role in controlling the entry of insecticides and other chemicals from the environment. Olson, W. suggests that the cuticular lipid layer constitutes a significant barrier to the penetration of lipoidal solutes through the integument. He concludes that partitioning of lipid-soluble insecticides such as DDT from the epicuticular lipid into lower phases is the rate-limiting step in penetration. Disruption of the cuticular lipids increases the rate of solute penetration into the cockroach. Other workers consider the underlying proteinacious part of the cuticle as the main barrier to insecticide penetration, and consider the surface lipid as facilitating the lateral movement of topically applied insecticides. Gerolt, P. has obtained evidence contradicting the widely held view that contact insecticides reach the site of action by penetrating through the integument. He suggests that topically applied dieldrin accumulates in the cuticular lipids, spreads laterally, and then reaches the site of action via the tracheal system.

The protection from micro-organisms provided by the cuticular lipids may be due to the non-wetting nature they provide the insect's surface, making it difficult for the organisms to establish themselves. It is also possible that the cuticular lipids provide chemical protection by being toxic to microorganisms. Koidsumi, K. reported that the cuticular lipids of the larvae of *Bombyx mori L.* and *Chilo simplex* Butler were capable of restricting the growth of pathogenic fungi *in vitro,* and indicated that the antifungal compounds were short-chain saturated fatty acids. However, no definitive structural identification of these compounds was made.

Chemical Communication

In addition to their critical role in preventing desiccation, insect cuticular lipids also serve in chemical communication. Major roles assigned to cuticular hydrocarbons include serving as sex pheromones, as species and caste recognition cues in several social insects, as thermoregulatory pheromones and as kairomonal cues for parasites.

The species reported to use cuticular components in chemical communication and lists the types of components involved.

Sex Pheromones

Cuticular hydrocarbons have been widely implicated as dipteran sex pheromones. Rogoff, W. *et al.* reported that benzene extracts of female *Musca domestica* elicited both attraction and excitation mating behavior patterns in males. Carlson, *D. et al.* reported that (Z)-9-tricosene was a "sex attractant" of *Musca domestica* and that it could be extracted from the bodies of females or from their fecal deposits.

A low level of response to (Z)-9-tricosene, both in olfactometer and pseudofly bioassays, prompted several workers to re-examine the female housefly's surface lipid chemistry, and led to the identification of over 100 isomeric methylalkanes, (Z)-9,10-epoxytricosane, (Z)-14tricosene-l0-one, and a C_{36} wax ester. The methylalkanes are terminally and internally branched monomethylalkanes and 3,X-, 4,X- and internally branched dimethylalkanes. Uebel, *E. et al.* showed that addition of branched alkanes of 28-30 carbons to a pseudofly treated with (Z)-9-tricosene enhanced the mating strike activity of the males, although the alkanes by themselves had no activity. This stimulatory effect of branched alkanes was confirmed by Rogoff, W. *et al.*. In addition, Uebel, E. *et al.* showed that the epoxide and ketone both released strike activity by the males in the absence of (Z)-9tricosene. The ester showed no biological activity. Combinations of all active and synergistic components still failed, however, to match a live virgin female as a target for a mating strike. Richter, I. *et al.* proposed that (Z)-9-tricosene may serve as a releaser mechanism for responsiveness to optical cues, leading to both mating strikes and aggregation.

Considerable effort has also been made at elucidating the importance of cuticular lipids as pheromones for the face-fly *Musca autumnalis*. Chaudhury, *M. et al.* suggested the presence of a sex pheromone in this species from olfactometer bioassay data. Uebel, *E. et al.* and Sonnet, *P. et al.* isolated and characterized several cuticular hydrocarbons which elicited strike behavior by males in a modified pseudofly bioassay. Active components included (Z)-14-nonacosene, (Z)-13-nonacosene and (Z)-13-heptacosene. These olefinic components were found in approximately equal quantities on both males and females. Males, however, contain greater quantities of heptacosane and nonacosane than do females, and the authors suggest that these alkanes attenuate the effects of the olefins. The overall hydrocarbon differences are however slight and a closer examination of this species likely to be fruitful.

An even more complex situation exists regarding the sex pheromone complex of the stable-fly, *Stomoxys calcitrans (L)*. Muhammed, *S. et al.* reported that the female produced both "sex attractant" and "mating stimulation" pheromones. Uebel, *E. et al.* showed that (Z)-9hentriacontene, (Z)-9-tritriacontene and methyl branched hentria- and tritriacontenes incited males to attempt copulation, as did a series of monomethyl- and dimethylalkanes. Later work by Sonnet, P. *et al.* on the saturated components showed that 15-methyl- and 15,19-dimethyltritriacontanes were most active. Harris, *R. et al.* showed that increasing amounts of these hydrocarbons were produced from emergence until the female was 3 days old. Meola, *R. et al.* reported that this species required a blood-meal for sex pheromone production and successful sexual behavior. Sonnet, *P. et al.* further characterized the female pheromone complex which induces male copulatory behavior. The major methyl branched olefins were identified as 13-methyl-lhentriacontene and 13-methyl- l-tritriacontene. The most active preparation was a combination of the normal alkenes, methylalkenes, and mono- and dimethylalkanes.

Three species of *Fannia* have been examined for sex pheromones using the pseudofly bioassay. (Z)9-Pentacosene constitutes the major cuticular hydrocarbon component of female *Fannia canicularis,* and was reported to stimulate the male to initiate copulatory attempts. The mating stimulants from female *Fannia pusio* are normal mono-olefins of 31 and 33 carbons. Bioassay data indicate that (Z)- 11 -hentriacontene is the most active component. Essentially the same mixture of alkenes is found on females of *Fannia femoralis*. For both species alkanes are also present, and Uebel, E. *et al.* showed that addition of the alkanes of *Fs femoralis* to its C_{31} olefins increased mating strike frequency.

Mackley, J. *et al.* and Bolton, *H. et al.* investigated the possibility that the horn-fly *(Haematobia irritans)* courtship behavior involved chemical components. Alkenes, including (Z)-9tricosene, (Z)-5-tricosene, (Z)-9-pentacosene and (Z)-9-heptacosene were identified from both sexes. Males had greater quantities of the (Z)-9-tricosene than did females, but the females contained greater quantities of all other olefins. Each alkane was individually active in releasing strike behavior, but only mixtures elicited the remaining steps of the courtship.

The fruit-fly *Drosophila melanogaster* also utilizes cuticular hydrocarbons as part of its sex pheromone. Newly emerged males and females have an almost identical hydrocarbon profile. A sexual dimorphism in long-chain hydrocarbons occurs starting on the second day after eclosion, and males lose sexual attractiveness. An aphrodisiac pheromone begins accumulating on females 1 day after eclosion. This hydrocarbon pheromone gives a dose-dependent wing vibration response in males.

Initial attraction of male tsetse flies to females is known to involve visual cues with no evidence presented for olfactory sex attractants. Good evidence does exist, however, for the presence of a contact sex pheromone in *Glossina morsitans* which elicits copulatory behavior by the males. Similar evidence exists for *Glossina palpalis palpalis* and *Glossina pallidipes* (Carlson, personal communication). Carlson, *D. et al.* found that male behavior associated with copulation was elicited only by three high-molecular weight hydrocarbons. The most active hydrocarbon was identified as 15,19,23-trimethylheptatriacontane, with 15,19-dimethylheptatriacontane and 17,21dimethylheptatriacontane being less active. Female *G. morsitans* was shown to contain about 4.2 *pg* of the trimethylalkane and 1.2 *pg* each of the dimethylalkanes. With all three compounds, male copulatory behavior was released only on physical contact with a surrogate female containing the chemicals. Huyton, P. *et al.* showed that solvent-extracted dead males baited with the trimethylalkanes gave an ED_{95} of 2 pg. No synergistic effects were found with the dimethylalkanes, and Huyton, P. *et al.* concluded that the trimethyl compound is the sex pheromone of this species. The natural pheromone appears on the pharate adult female about 2 days before emergence from the puparium and is present throughout her life. Male responsiveness to pheromone baited decoys increases to a maximum 3-4 days after emergence. These authors suggest that another possible function of this pheromone may be to ensure that the union of male and female is maintained throughout the lengthy copulation period which is required for successful ejaculation and spermatophore transfer.

Huyton, *P. et al.* conducted interspecific mating tests with seven species of *Glossina* to evaluate the specificity of the contact pheromone isolated from *G. morsitans*. Their tests with either live or dead females as targets indicated that the responses of the males are largely the consequence of differences in composition of the female surface cuticular hydrocarbons and not of female behavior.

In addition to Diptera, other insects use longchain hydrocarbons as sex pheromones. Several sibling species in the genus *Holomelina* (tiger moths) use 2-methylheptadecane, as do females of *Pyrrharctia isabella*. Conner, W. *et al.* noted that the female arctiid moth *Utetheisa ornatrix* emits (Z,Z,Z)-3,6,9heneicosatriene plus small quantities of an unidentified C_{21} tetraene as her sex pheromone. These hydrocarbons appear gland-specific and are probably not cuticular components.

TABLE 5.1: CUTICULAR LIPIDS IN CHEMICAL COMMUNICATION

Process/species	Components involved
Sex pheromone	
Musca domestica	(Z)-9-tricosene, (Z)-9, 10-epoxytricosane (Z)-14-tricosene-10-one, methylalkanes *Musca autumnalis* (Z)-14-nonacosene, (Z)-13-nonacosene, (Z)-13-heptacosene
Stomoxys calcitrans	(Z)-9-hentriacontene, (Z)-9-tritriacontene, methylalkanes 15-methyl and 15, 19-dimethyltritriacontanes 13-methyl-l-hentriacontene, 13-methyl-l-tritriacontene
Fannia canicularis	(Z)-9-pentacosene + alkanes
Fannia pusio	(Z)-11-hentriacontene + alkanes
Fannia femoralis	(Z)-11-hentriacontene + alkanes
Haematobia irritans	(Z)-5-tricosene, (Z)-9-pentacosene, (Z)-9-heptacosene
Glossina morsitans	15, 19, 23-trimethyheptatriacontane, 15, 19- and 17, 21-dimethylheptatriacontane *Glossina sp.* methylalkanes (seven other species)
Lucilia cuprina	cuticular lipids (unidentified)
Pikonema alaskensis	9, 19-alkadienes of C_{29}, C_{31}, C_{33}, C_{35} and C_{37}
Lycoriella mali	*n-heptadecane*
Colias eurytheme	13-methylheptacosane *and* esters
Copulation release pheromone	
Callosobruchus chinensis	*mono- and* dimethylalkanes of C_{26}-C_{33} and (E)-3, 7-dimethyl-2-octene-1, 8-dioic acid
Species and caste recognition	
Zooptermopsis angusticollis	hydrocarbon composition
Reticulitermes flavipes	hydrocarbon composition
Reticulitermes virginicus	hydrocarbon composition
Trichopsenius frosti	hydrocarbon composition
Trichopsenius depressus	hydrocarbon composition

Philotermes	*howardi* hydrocarbon composition
Xenitusa hexagonalis	hydrocarbon composition
Kairomones	
Heliothis zea (attract parasites)	13-methylhentriacontane
Heliothis virescens (attract parasites)	11-methylhentriacontane, 16-methyldotriacontane, 13-methyltriacontane
Thermoregulatory	
Vespa crabro	(Z)-9-pentacosene
Oviposition markers	
Cardiochiles nigriceps	nC_{23}—nC_{33}

Two sulfur butterflies, *Colias eurytheme* and *Colias philodice* (Pieridae) coexist over much of North America and originally were thought to interbreed. Reproductive isolation is based upon female response to species-specific visual and chemical signals. Males of *C. philodice* contain on their wings three major esters (n-hexylmyristate, nhexylpalmitate, and n-hexylstearate), a series of *n*-alkanes with odd carbon numbers (C_{23} to C_{29}), and small quantities of 13-methylheptacosane. Males of *C. eurytheme* contain on their wings the same *n*alkanes, no esters, and a large amount of 13-methylheptacosane. The esters and 13-methylheptacosane all have significant electrophysiological activity. Preliminary behavioral data suggest that the esters are the important species recognition cues and that the branched hydrocarbon functions as an aphrodisiac. Grula, J. and Taylor, O. studied the inheritance of male pheromone production by gas chromatographic analysis of wing extracts of various genotypes derived from forced interspecific crosses. The X-chromosome carried most of the gene(s) controlling production of 13-methylheptacosane. Expression of this compound in hybrids displayed a codominant pattern, and was not influenced by diet.

Cuticular hydrocarbon components are part of the sex pheromone of the yellowheaded spruce sawfly, *Pikonema alaskensis* Rohwer. A series of (Z,Z)-9,19-alkadienes of 29, 31, 33, 35 and 37 carbons are present on the surface of the female and possess pheromone activity.

Long-chain methyl branched alkanes are components of the copulation release pheromone, erectin, of the azuki bean weevil, *Callosobruchus chinensis L.*. 3-Methyl-, internally branched monomethyl- and dimethylalkanes of C_{29} C_{35}along with a dicarboxylic acid comprise the copulation release pheromone, which is distinct from the sex attractant.

Species and Caste Recognition Cues

Social insects are well known for their ability to recognize conspecifics, and to distinguish the caste and sex of the individuals they interact with. Because many social insects live either wholly in the soil or distributed between soil and wood (especially subterranean termites and many ants), much of their sensory perception must be either olfactory or tactile. The nests and galleries of these insects are often tightly closed, low-volume systems. Such systems place a premium on semiochemicals which are complex enough to have a high information content, but which are of low enough volatility to minimize sensory habituation. Howard, R. *et al.* and Blomquist, G. *et al.* suggested that the cuticular hydrocarbons of termites might serve as semiochemical cues for caste and species recognition. They found that *Reticulitermes flavipes* (Rhinotermitidae) and *Zootermopsis angusticollis* (Hodotermitidae) possess drastically different hydrocarbon profiles, and both of these differ markedly from *Nasutitermes exitiosus* (Nasutitermitidae). In addition, R. Howard and co-workers completely characterized the cuticular hydrocarbons of *R. virginicus* and partially characterized by GLC the hydrocarbons of four other species of *Reticulitermes,* as well as those of *Coptotermes formosanus, Marginitermes hubbardi, Incistitermes minor, Pterotermes occidentis,* and *Tenuirostitermes tenuirostris* (unpublished data). In every case, the hydrocarbon composition is unique. Furthermore, examination of the hydrocarbon compositions by caste indicates castespecific ratios of hydrocarbon components. Preliminary bioassay data have confirmed the role of cuticular hydrocarbons as species recognition cues for *Reticulitermes virginicus* and *Reticulitermes flavipes*.

Additional evidence that cuticular hydrocarbons serve as species recognition cues comes from the finding that the highly integrated, host-specific termitophilous beetle *Trichopsenius frosti* has an identical cuticular hydrocarbon profile as its host termite *R. flavipes.* In addition, the cuticular hydrocarbons of three species of termitophilous beetles (representing two subfamilies of Staphylinidae) associated with *R. virginicus* are identical to those of that termite (Howard, unpublished data). Biosynthetic studies showed that *T. frosti* produces its own hydrocarbons, rather than procuring them in some manner from its termite host.

The scarab beetle *Myrmecaphodius excavaticollis* associated with *Solenopsis spp.* ("fire ants") has a cuticular hydrocarbon composition that loosely mimics that of its current ant host. The mechanism by which the beetles achieve this is unknown. The cuticular hydrocarbons of two *Solenopsis* species *(Solenopsis invicta* and *Solenopsis richteri)* have been thoroughly characterized and those of two other species *(Solenopsis geminata* and *Solenopsis xyloni)* characterized by GLC (Vander Meer, personal communication). All four species have unique hydrocarbon profiles, suggesting that, like termites, ants may use these chemicals as species recognition cues.

Thermoregulatory Pheromones

Wasps and hornets are known to actively heat their brood combs, especially those cells containing pupae. They do this by positioning their respiratory spiracles over the silken dome of the pupal cell and alternately retracting and extending their abdominal segments up to 180 times per minute. Ishay showed by preliminary bioassays that it was the pupae themselves that triggered this brooding behavior, and that the stimulus involved was that of a "volatile" chemical which could be extracted with alcohol from the pupae. Veith, H. and Koeniger, N. (1978) identified the pheromone as (Z)-9-pentacosene.

Kairomonal Cues for Parasites

Parasites are such ubiquitous components of practically every ecosytem that it is not surprising that some of them have managed to evolve to utilize cuticular hydrocarbon cues produced by their hosts as primary cues in locating their hosts. Hymenopteran parasites of *Heliothis* spp. exhibit intensified searching behavior for their hosts when exposed to either frass or scales of the ovipositing female moth. Jones, *R. et al.* showed that 13methylhentriacontane was the major component of the hemolymph, cuticle, and frass of larvae *Heliothis zea* which was responsible for intense searching behavior by the larval parasite *Microplitis croceipes*. Vinson, *S. et al.* similarly showed that 11-methylhentriacontane, 16-methyldotriacontane and 13-methyltritriacontane, produced in the mandibular glands of *Heliothis virescens* and deposited on the leaf surface by feeding larvae, stimulate *Cardiochiles nigriceps* to search the immediate vicinity more efficiently. Jones, *R. et al.* reported that an egg parasite of *H. zea (Trichogramma evanescens)* was stimulated by extracts of scales left by the ovipositing *H. zea* female, and that the most active factor in these extracts was n-tricosane. Lewis and co-workers have since conducted extensive laboratory and field investigations on the role and practical utilization of n-tricosane in manipulating population levels of *H. zea*. Much of their findings regarding the behavioral sequence of host selection by parasites, the chemistry of the kairomones involved, their role in the host finding process, and the basis for their potential employment in pest management programs has recently been reviewed.

Wax of Honeybees, Scale Insects

In addition to a thin layer of cuticular lipid, some insects secrete large amounts of wax. These include the honeybees, which use the wax as structural material to build the honeycomb, and the scale insects, which use the wax for protection. The scale insects are so called because in many species the female is protected by a scale or shield consisting of a mixture of wax and cast skins. The amount of lipid covering the scale insects is much larger than that usually seen on the epicuticle of most insects; however, its chief function is still most likely to protect the insects against desiccation. The adult females of scale insects are degenerate and are attached or fastened to the host plant by the mouthparts. It is probably because they are immobile that the scale insects form their characteristic protective coat, which also protects them from insect predators and the weather.

ANALYTICAL PROCEDURES

Extraction and Separation

The analytical techniques used in insect wax chemistry have been reviewed comprehensively by Jackson, L. and Armold, M. and by Gilby, A.. The cuticular lipids from whole insects or cast exuviae can be conveniently extracted by immersion in organic solvents such as hexane or chloroform for short periods of time. Usually a few seconds to 15 min is recommended for hexane, and shorter extraction times for chloroform. Chloroform is more penetrating and the chances of extracting internal lipids increase as the extraction time increases. Chloroform-methanol mixtures and diethylether are not commonly used except for very short extraction periods due to their interaction with water and the increased possibility of extracting internal lipids. Most often one immersion does not extract all of the cuticular lipid; two or more short immersions are usually better than one long immersion.

After extraction, the major lipid classes can usually be separated by column chromatography and/or thin-layer chromatography (TLC). The use of silver nitrate-impregnated TLC plates for unsaturated components and molecular sieve for branched components are often very helpful. Gas-liquid chromatography (GLC) is routinely used to separate and quantify the components of each class of lipid. Infrared spectrophotometry, nuclear magnetic resonance spectrometry, mass spectrometry (MS) and combined GLC-MS, including chemical ionization (CI)-MS techniques have proven very useful in structure determination. GLC-MS is the principal tool used in obtaining structure assignments for hydrocarbons.

Structure Determination of Hydrocarbons

n-Alkanes can be identified by their retention times on GLC and inclusion in 5 A molecular sieve. Structure assignments can be confirmed by GLC-MS.

The methyl branched alkanes have retention times on GLC which are somewhat less than an nalkane of the same total carbons. Monomethylalkanes with methyl branches on carbons 2, 3 or 4 elute about 0.3 or 0.4 carbon units in front of nalkanes of the same total carbon number on packed columns. Internally branched monomethylalkanes elute about 0.6 to 0.7 carbon units in front of corresponding n-alkanes. Internally branched dimethylalkanes elute about 1.3 to 1.4 carbon units before corresponding n-alkanes, and dimethlyalkanes with one methyl branch near the end and one in the middle (3,X-, 4,X-) elute about 1 carbon unit in front of corresponding n-alkanes.

When subjected to mass spectrometry each type of hydrocarbon yields a diagnostic spectrum. Discussions on the interpretation of spectra from insect hydrocarbons can be found in papers by Jackson, L. and Blomquist, G., Nelson, D. and Sukkestad, D., Nelson, *D. et al.*, Nelson, D., and Pomonis, J. *et al.*. Recently, the application of chemical ionizationmass spectrometry to insect hydrocarbons has proven useful.

Most of the alkenes characterized from insects to date are *cisl(Z)*. This assignment is usually based on the presence of a peak at 730 cm $^{-1}$ and the absence of a peak at 970 cm^{-1} (indicative of *trans* or E double bonds) in the infrared spectrum. Z and E isomers can be separated by silver nitrate-impregnated TLC. Oxidative cleavage followed by GLC or oxidation of the alkene to a diol with osmium tetroxide followed by silylation and mass spectrometry have been successfully used to determine the position of double bonds. More recently, methoxymercuration-demercuration followed by mass spectral analysis has been successfully applied to alkenes. This technique is especially advantageous with mixtures of components, as it eliminates the need for isolating each component by preparative GLC.

Characterization of Wax Esters, Fatty Alcohols and Sterols

Wax esters are usually characterized by saponification and characterization of the component fatty acids and fatty alcohols.

However, some information can be obtained about the total number of carbon atoms in the wax ester chains by GLC on a high-temperature silicone column. Since GLC only gives peaks according to total number of carbons in the wax ester there may be a mixture of various alcohols esterified to a mixture of fatty acids in any one peak. In order to isolate the alcohols and fatty acids, the wax esters are usually either saponified or transesterified. Saponification has the advantage that the alcohols can be extracted with an organic solvent from the alkaline saponification mixture and then extraction of the acidified solution will yield the fatty acids. Transesterification provides the methyl esters of the fatty acids which are easily separated from the alcohols by TLC or column chromatography and then the fatty acid methyl esters and fatty alcohols can be readily characterized by GLC. Free primary alcohols, secondary alcohols and sterols can usually be separated from the other lipid components by TLC or column chromatography. The purified fractions of the various alcohols are then submitted to GLC and GLC-MS for further characterization.

Secondary alcohol wax esters require some special techniques since they do not saponify or transesterify well under normal conditions. Secondary alcohol wax esters have been characterized by GLC of the total fraction to determine the range of chain lengths. This step is followed by $LiAlH_4$ reduction of the ester linkage to yield secondary alcohols and primary alcohols. TLC separation of primary and secondary alcohols followed by GLC or GLC-MS can be used to characterize the components of the secondary alcohol wax esters.

Characterization of Free Fatty Acids and Acylglycerols

The free fatty acid fraction is generally characterized by methylation followed by GLC on a polyester column which separates the methylesters according to chain length and number of double bonds. The triacylglycerols, diacylglycerols and monoacylglycerols can be separated by TLC or column chromatography and then each fraction can be analyzed by GLC on high-temperature silicone columns, or they can be simply saponified and methylated or transesterified and the resulting methylesters can be subjected to GLC analysis.

STRUCTURE AND COMPOSITION OF CUTICULAR LIPIDS

The cuticular lipids of many insect species consist of complex mixtures of various types of hydrocarbons and other components.

<u>n</u>-alkanes	$CH_3-(CH_2)_x-CH_3$
<u>n</u>-alkenes	$CH_3-(CH_2)_x-CH{=}CH-(CH_2)_y-CH_3$
2-methylalkanes	$CH_3-\overset{CH_3}{\overset{\vert}{C}H}-(CH_2)_x-CH_3$
3-methylalkanes	$CH_3-CH_2-\overset{CH_3}{\overset{\vert}{C}H}-(CH_2)_y-CH_3$
internally branched monomethylakanes	$CH_3-(CH_2)_x-\overset{CH_3}{\overset{\vert}{C}H}-(CH_2)_y-CH_3$
dimethylalkanes	$CH_3-(CH_2)_x-\overset{CH_3}{\overset{\vert}{C}H}-(CH_2)_y-\overset{CH_3}{\overset{\vert}{C}H}-(CH_2)_z-CH_3$
trimethylalkanes	$CH_3-(CH_2)_y-[\overset{CH_3}{\overset{\vert}{C}H}-(CH_2)_y]_3-(CH_2)_z-CH_3$
primary alcohol wax esters	$CH_3-(CH_2)_x-O-\overset{O}{\overset{\Vert}{C}}-(CH_2)_y-CH_3$
secondary alcohol wax esters	$CH_3-(CH_2)_x-\underset{(CH_2)_y-CH_3}{\underset{\vert}{\overset{H}{\overset{\vert}{C}}}}-O-\overset{O}{\overset{\Vert}{C}}-(CH_2)_z-CH_3$
epoxides	$CH_3-(CH_2)_x-\overset{O}{CH-CH}-(CH_2)_y-CH_3$
ketones	$CH_3-(CH_2)_x-\overset{O}{\overset{\Vert}{C}}-(CH_2)_y-CH_3$
secondary alcohols	$CH_3-(CH_2)_x-\overset{OH}{\overset{\vert}{C}H}-(CH_2)_y-CH_3$
primary alcohols	$CH_3-(CH_2)_x-CH_2-OH$
free fatty acids	$CH_3-(CH_2)_x-\overset{O}{\overset{\Vert}{C}}-OH$
sterols (cholesterol)	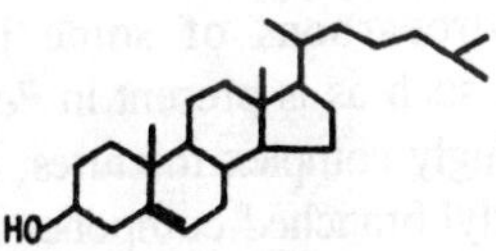 HO

Fig. 5.1. Structures of common insect cuticular compoents.

Hydrocarbons

Hydrocarbons have been the most extensively studied cuticular components and frequently are the only class characterized. They are common components of surface extracts and are present on all insects which have been carefully examined. The insect surface lipids of some species are > 90% hydrocarbon whereas hydrocarbons account for only 0.5% of the surface lipid of the tobacco budworm pupae. Insect surface lipid hydrocarbons are usually a mixture of components which includes n-alkanes, n-alkenes, terminally branched monomethylalkanes, internally branched monomethylalkanes, dimethylalkanes, trimethylalkanes and others.

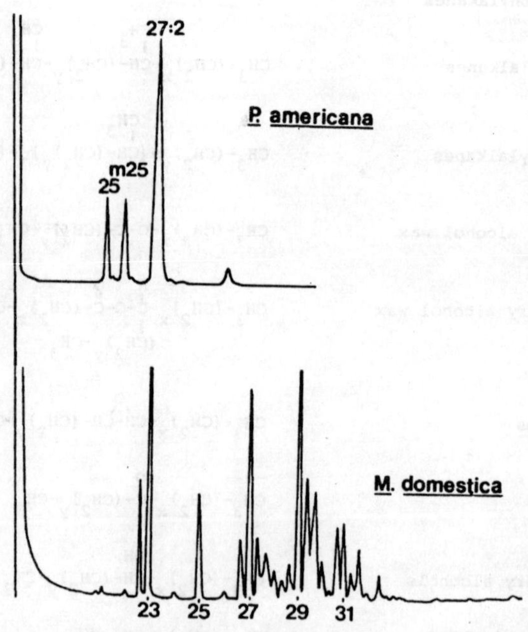

Fig. 5.2. GLC traces of the cuticular hydrocarbons of Periplaneta americana and Musca domes tica. The components in the top trace are 25(n-pentacosane), m25(3-methylpentacosane), and 27: 2 (IZ,ZI-6,9-heptacosadiene). The numbers below the bottom trace refer to the carbon numbers of n-alkanes.

The hydrocarbons of some insects have a relatively simple composition, such as is present in *Periplaneta americana.* Other insects have exceedingly complex mixtures, such as the female housefly, which has 134 methyl branched components.

The complexity of insect cuticular hydrocarbons has led to the suggestion that hydrocarbon composition might be used as a taxonomic character. However, only a relatively small number of species have been examined and there do not appear to be obvious correlations of hydrocarbon compositions within insect groups. In addition, the composition of hydrocarbons often varies tremendously among different developmental forms, between sexes, and at different times of the year. In general, most investigators have concentrated on one developmental stage. These observations indicate that the distribution pattern of hydrocarbon components is unlikely to be a useful discriminating factor in the near future.

n-Alkanes

n-Alkanes occur in all the insect surface lipids so far investigated. The n-alkanes of insect surface lipids are generally in the range C_{21}-C_{36} with alkanes having an odd number of carbon atoms predominating. The complexity of the n-alkane mixtures varies considerably. For example, n-pentacosane comprises 85% of the n-alkane fraction of *P. americana*, but *n*-alkanes from C_{17} to C_{34} with both odd and even numbers of carbon atoms are major components of the hydrocarbons of pupal tobacco budworms, *Heliothis virescens*. *n*-Alkanes are not usually found alone, but only n-alkanes were reported from the hydrocarbon fraction from cast skins of the beetle, *Tenebrio molitor*. Analysis of the hydrocarbons of seven scale insects showed that they were entirely n-alkanes of C_{25}-$C_{35,}$ with odd-numbered carbon chain-lengths predominating.

Terminally Branched Monomethylalkanes

2-Methylalkanes and/or 3-methylalkanes are prevalent in insect surface lipids. Like the n-alkanes, the terminally branched monomethylalkanes range from simple compositions such as in *P. americana* where only 3-methylpentacosane is present to complex mixtures of 3methylalkanes, 2-methylalkanes or mixtures of 2and 3-methylalkanes. The 2-methylalkanes are somewhat unique in that components with both odd- and evennumbered carbon chains are present in substantial amounts. This appears to reflect their biosynthesis, which involves the carbon skeleton of either valine or leucine. Many of the insects studied to date that have internally branched monomethylalkanes also have 3-methylalkanes, which may reflect a similar biosynthetic origin. The majority of the 3-methylalkanes have an odd-numbered carbon chain (even number of total carbons).

Internally Branched mono-, di- and Trimethylalkanes

A very comprehensive review covers the occurrence and analysis of internally methyl-branched hydrocarbons. The majority of the internally branched monomethylalkanes have the methyl branch located on an odd-numbered carbon atom which often is carbon number 5, 7, 9, 11, 13, 15 or 17. The general range is from 20 to 40 carbons. Monomethylalkanes with the methyl branch on an even-numbered carbon atom are usually reported mixed with odd-numbered carbon isomers and as minor components.

Most of the long-chain internally branched dimethylalkanes have an isoprenoid type spacing, although they are not derived from isoprenoid units. Common isomers are 9,13-, 11,15-, 13,17-, 15,19- and 17,21-. Methyl branches on positions 11,21- have been reported and an unusual symmetrical component, 5,17dimethylheneicosane, has been reported from the termite *Zootermopsis angusticollis.* Other types of dimethylalkanes include a 3,X- and 4,X-dimethylalkane series from the housefly and cigarette beetle, 11,19-dimethylalkanes in the tobacco budworm, 11,12-dimethylalkanes in the cigarette beetle and 13,15dimethylalkane in the fire ant. As careful analyses of dimethylalkanes are made on more organisms, it appears that the methyl groups can be positioned almost anywhere on the chain. Isomers with methyl groups on oddnumbered carbons are more prevalent than isomers with methyl branches on even-numbered carbons. Internally branched trimethylalkanes with isoprenoid spacing of the methyl branches have been reported from *Manduca sexta*, *Schistocerca vaga* and the tsetse fly *Glossina morsitans.*

Two homologous series of trimethylalkanes have been observed as major constituents of the saturated hydrocarbons from ants (Martin, M. and MacConnell, J., 1970). 3,7,1 1-Trimethylalkanes of C_{34}, C_{36} and C_{38} and 4,8,12-trimethylalkanes of C_{35}, C_{37} and C_{39} are the major constituents of the surface lipid-saturated hydrocarbons of the ant, *Atta columbia*. In *Atta sexdens,* both 3,7,11- and 4,8,12-trimethylalkanes are present as major components, but in *Atta cephalotes isthmicola,* only the 3,7,11-trimethyl series is present and accounts for only about 15% of the alkanes.

Alkenes

n-Alkenes, with one, two, or three double bonds, have been characterized from about one-half of the insects examined to date. The chain length of cuticular alkenes usually ranges from C_{20} to C_{37}, with odd-numbered chain lengths predominating. The positions of the double bonds can be almost anywhere in the chain. Alkenes with the double bond in the 1-position are present in the red flour beetle,

Tribolium castaneum. Two and 3-alkenes, as well as a series of alkenes with double bonds in unspecified positions, ranging from C_{20} to C_{31}, have been reported in the pecan weevil *Curculio caryae.* A number of Dipteran species possess *n*-alkenes of 23 to 31 carbons which function in chemical communication. The double bonds are usually either in the 7 or 9 positions. In three cockroaches, *Periplaneta australiasia, Periplaneta fuliginosa* and *Periplaneta japonica,* (Z)-9-alkenes from C_{27} to C_{31} were observed. All castes of the termites *Reticulitermes flavipes* and *Reticulitermes virginicus* possess (Z)-9pentacosene as a major cuticular component as do four species of termitophilous beetles associated with them. Alkenes with double bond positions at 8 and 10 have been reported from the honeybee, *Apis mellifera.* The face-fly, *Musca autumnalis,* has a complex series of 6-, 8-, 10-, 11- and 12-pentacosene, 10-, 11-, 12- and 13heptacosene, and 10-, 11-, 12-, 13-, and 14nonacosene.

Polyunsaturated hydrocarbons are not as common insect surface lipid components and are often minor components. However, the major hydrocarbon component of the surface lipids of the American cockroach, *Periplaneta americana,* is (Z,Z)-6,9-heptacosadiene.

Although most of the other cockroaches studied have methylalkanes as the major components, P. *japonica* hydrocarbons are 17% (Z,Z)-6,9nonacosadiene and lesser quantities of (Z,Z)-6,9heptacosadiene, -octacosadiene, -tricontadiene and hentriacontadiene. Cuticular dienes having chain lengths of C_{29}-C_{37} with double bonds in the (Z,Z)-9,19-positions are sex pheromone components of the female yellowheaded spruce sawfly, *Pikonema alaskensis.*

OXYGENATED COMPONENTS

The non-hydrocarbon components of insect surface lipids have not been extensively studied in many insects, but it appears that wax esters seldom predominate except in beeswax. Free alcohols, free fatty acids and lesser quantities of other components are common. Some insects have appreciable quantities of unusual lipids and some of those will be discussed below.

Wax Esters and Sterol Esters

Wax esters are not as common in insect cuticular lipids, nor have they been as widely studied, as hydrocarbons. Wax esters from different insects vary in composition, although most of them are composed of long-chain normal primary alcohols and acids. The cuticular lipids of both the adult and naiad *Pteronarcys californica* contain saturated and unsaturated C_{30} to C_{40} wax esters but the adults have more of the longer-chain wax esters. The pea aphid, *Acyrthosiphon pisum,* contains C_{32} to C_{54} aliphatic wax esters composed of C_{12} to C_{34} normal saturated primary alcohols and acids. Sterol esters are also reported in the cuticular lipids of some insects with the principal sterol being cholesterol.

Novel wax esters were identified in the cuticular lipids of the grasshoppers *Melanoplus sanguinipes* and *Melanoplus packardii.* These wax esters, comprising 18 and 28% respectively of the cuticular lipids of these two insects, consist of saturated fatty acids esterified to secondary alcohols. The fatty acids range from C_{12} to C_{22}. The secondary alcohols range from C_{21} to C_{27} with the hydroxyl group near the center of the carbon chain. In *M. sanguinipes,* 59% of the secondary alcohols consist of the C_{23} compounds in which the major isomer is tricosan-11-ol, with smaller amounts of tricosan-l2-ol and tricosan-l0-ol. A number of other species of *Melanoplus* also contain these unusual wax esters, and to date they have not been reported from other insects.

The primary alcohols found in the cuticular lipids of insects are usually of even-numbered carbon atoms and range from C_{22} to C_{34}. Seventy percent of the cuticular lipid of the larva of the eri silkworm, *Samia cynthia ricini* is *n*-triacontanol and n-octacosanol. The alcohols from *Eleodes armata* are primarily saturated C_{20} C_{34} alcohols with the suggestion of branched and/or unsaturated alcohols containing 30-40 carbon atoms. Over 90% of the wax on *Eriocampa ovata* larvae is primary alcohol of C_{21}—C_{12}, with hexacosanol comprising about 73% of the fraction (Percy, and Blomquist, unpublished data). In *M. sanguinipes* and *M. packardii* the primary alcohols are C_{24} C_{30} and comprise only a small percentage of the surface lipids. Trace amounts of free secondary alcohols are also present from *M. sanguinipes* and *M. packardii*. Half of the surface lipid extract of larvae of *Tenebrio molitor is* pentacosan-8,9-diol with a melting point of 115°. A distinction between insects with "soft" surface lipids in which the proportion of hydrocarbons is high, and those with "hard" surface lipids in which long-chain alcohols (and diols) appear to predominate, is suggested.

Fatty Acids

The C_{14}-C_{20} fatty acids with an even number of carbon atoms predominate. In most cases saturated and unsaturated fatty acids have been observed. However, the surface lipid fatty acids of *Lucilia cuprina* are all saturated. Branched fatty acids are not usually present in insect waxes. Free a-hydroxy acids have been observed from the pea aphid, *Acyrthosiphon pisum.*

Other Non-Hydrocarbon Cuticular Components

The three major non-hydrocarbon components of the surface lipids of female houseflies are (Z)-14tricosene-10-one, (Z)-9,10-epoxytricosane and 9hexadecenyl-9-octadecenoate. The first two compounds may be involved in the sex pheromone complex. The surface lipid of male *Drosophila melanogaster* contain (Z)-vaccenyl acetate as a major nonhydrocarbon component. The (Z)-vaccenyl acetate is found in the

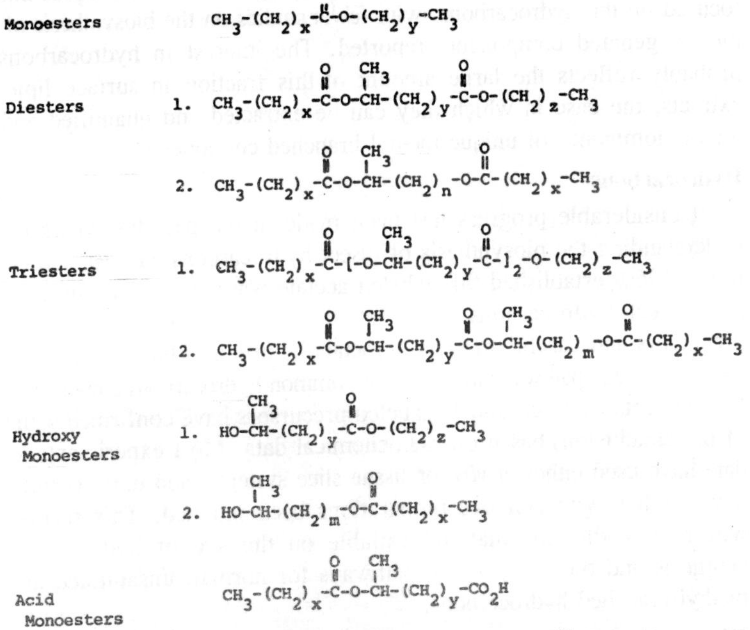

Monoesters $CH_3-(CH_2)_x-\overset{O}{\overset{\|}{C}}-O-(CH_2)_y-CH_3$

Diesters

1. $CH_3-(CH_2)_x-\overset{O}{\overset{\|}{C}}-O-\overset{CH_3}{\overset{|}{CH}}-(CH_2)_y-\overset{O}{\overset{\|}{C}}-O-(CH_2)_z-CH_3$

2. $CH_3-(CH_2)_x-\overset{O}{\overset{\|}{C}}-O-\overset{CH_3}{\overset{|}{CH}}-(CH_2)_n-O-\overset{O}{\overset{\|}{C}}-(CH_2)_x-CH_3$

Triesters

1. $CH_3-(CH_2)_x-\overset{O}{\overset{\|}{C}}-[-O-\overset{CH_3}{\overset{|}{CH}}-(CH_2)_y-\overset{O}{\overset{\|}{C}}-]_2-O-(CH_2)_z-CH_3$

2. $CH_3-(CH_2)_x-\overset{O}{\overset{\|}{C}}-O-\overset{CH_3}{\overset{|}{CH}}-(CH_2)_y-\overset{O}{\overset{\|}{C}}-O-\overset{CH_3}{\overset{|}{CH}}-(CH_2)_m-O-\overset{O}{\overset{\|}{C}}-(CH_2)_x-CH_3$

Hydroxy Monoesters

1. $HO-\overset{CH_3}{\overset{|}{CH}}-(CH_2)_y-\overset{O}{\overset{\|}{C}}-O-(CH_2)_z-CH_3$

2. $HO-\overset{CH_3}{\overset{|}{CH}}-(CH_2)_m-O-\overset{O}{\overset{\|}{C}}-(CH_2)_x-CH_3$

Acid Monoesters $CH_3-(CH_2)_x-\overset{O}{\overset{\|}{C}}-O-\overset{CH_3}{\overset{|}{CH}}-(CH_2)_y-CO_2H$

Fig. 5.3. Structures of the major ester components of beeswax.

ejaculatory bulb of adult males and is transferred to females during mating. Heneicosan-8-ol acetate was identified as a major (27%) nonhydrocarbon constituent of 5-day-old male littlehouseflies, *Fannia canicularis*. Triacylglycerols are present in some cuticular extracts,, but usually as minor amounts. Their presence in extracts of cast skins would suggest that the triacylglycerols are true cuticular components, and not simply contaminants from internal lipids.

Beeswax-Surface Lipids and Comb Wax

The composition of the surface lipid of the honeybee, *Apis mellifera, is* quantitatively different from that of the comb wax. The major component of the surface lipid is hydrocarbon (58%), followed by monoesters (23%), diesters (9%), triesters (2%), free fatty acids and more polar lipids (8%) (Blomquist, G. *et al.*, 1980b). The comb wax composition is hydrocarbon (13-17%), wax monoester (31-35%), diesters (10-14%), triesters (3%), hydroxymonoesters (3-6%) and hydroxypolyesters, free primary alcohols, diols, and acid monoesters (7-10%).

BIOSYNTHESIS OF CUTICULAR LIPIDS

Most of the work on the biosynthesis of insect cuticular lipids has focused on the hydrocarbons, with fewer studies on the biosynthesis of the oxygenated components reported. The interest in hydrocarbons probably reflects the large amount of this fraction in surface lipid extracts, the ease in which they can be extracted and quantified and the predominance of unique methyl branched components.

Hydrocarbons

Considerable progress has been made in the past few years on understanding the biosynthesis of insect hydrocarbons. *In vivo* studies in the 1960s established that labeled acetate was readily incorporated into insect hydrocarbons. Later studies with specific radiolabeled precursors and careful analysis of metabolic products have established the biosynthetic pathways for the most common hydrocarbon components. Experiments using carbon-13 labeled precursors have confirmed some of the conclusions based on radiochemical data. Most experiments to date have used either *in vivo* or tissue slice systems, and only recently have studies with cell-free preparations been initiated. This section will review the information available on the site of hydrocarbon synthesis and the biosynthetic pathways for normal, unsaturated and methyl branched hydrocarbons.

Dietary Contribution

In a study comparing the cuticular hydrocarbons of the tobacco hornworm to its dietary hydrocarbons, Nelson, D. *et al.* presented data indicating that a portion of the dietary alkanes was incorporated into the cuticular lipids. A similar conclusion was reached by Blomquist, G. and Jackson, L. using radiolabeled alkanes fed to a grasshopper. However, in both of these cases the contribution of dietary alkanes to the insect's complement of hydrocarbon appeared small, and the insect biosynthesized the majority of its hydrocarbon components. In a study on the source of the hydrocarbons which serve as kairomones of the corn earworm, Hendry, *L. et al.* concluded that the diet was the major source of the kairomone (ntricosane). However, they did not examine the ability of the corn earworm to produce *n*-tricosane *de novo*. In as much as many other insects are known to do so readily, it would seem wise to check this point before excluding the possibility.

Site of Synthesis

The most probable site(s) for cuticular hydrocarbon biosynthesis are cells associated with the epidermal layer or the peripheral fat body. Nelson, D. demonstrated that acetate and palmitate were incorporated *in vitro* into hydrocarbons by the integument of *Periplaneta americana* and *Manduca sexta,* whereas the fat body did not efficiently incorporate either of these precursors into hydrocarbon. Similarly, studies have demonstrated that hydrocarbon production in *Periplaneta americana, Sarcophaga bullata* and *Musca domestica* occurs in epidermal related cells. Diehl, P. separated the oenocyte-rich peripheral fat body from the central yellow fat body tissue in the locust, *Schistocerca gregaria,* and observed the highest rate of hydrocarbon synthesis in the oenocyte-rich peripheral fat body. In *Tenebrio molitor,* Romer, F. demonstrated that isolated oenocytes efficiently and specifically incorporated [1-^{14}C]acetate into hydrocarbon.

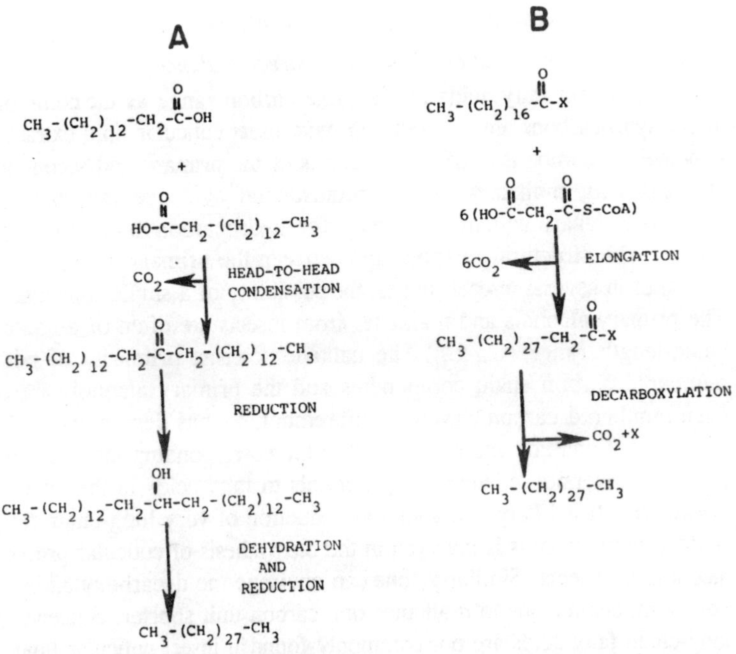

Fig. 5.4. The two most considered pathways for hydrocarbon biosynthesis. A: Head-to-head condensation elongation-decarboxylation pathway.

Transport

Chino and co-workers have recently shown that the diacylglycerol-carrying lipoprotein (DGLP) (or lipophorin) from the American cockroach and a locust contain large amounts of hydrocarbon. Approximately one-fourth of the total lipids associated with lipophorin were hydrocarbon which were similar in composition to those found on the surface of the insects. They postulate that one function of lipophorin is to carry hydrocarbon from the site of synthesis to the site of deposition.

Biosynthetic Pathways

The two most considered metabolic routes for hydrocarbon biosynthesis involve condensationreduction and elongation-decarboxylation pathways. In the bacterium, *Sarcinia lutea,* evidence has been presented favoring a modification of the condensation-reduction pathway, in which two acyl groups condense head-to-head with the loss of a carboxyl group, and the molecule is subsequently reduced. Incontrast, Kolattukudy and co-workers have presented convincing evidence that in plants hydrocarbon biosynthesis occurs by the elongation of fatty acids followed by a reductive decarboxylation. Both indirect and direct evidence favors the elongation-decarboxylation pathway in insects.

Elongation-decarboxylation pathway: indirect evidence

Long chain fatty acids of the same carbon range as the common insect hydrocarbons are not found in most insect cuticular lipid extracts. However, a comparison of the structures of the primary and secondary alcohols with n-alkanes lends circumstantial evidence favoring the elongation-decarboxylation pathway for hydrocarbon biosynthesis in insects. The structural relationships between the primary alcohols and n-alkanes in several insects suggest the possibility of a similar precursor. The primary alcohols and n-alkanes from insects are often of a similar chain-length range (C_{22}-C_{34}). The nalkanes consist primarily of odd-numbered carbon chain components and the primary alcohols are of even-numbered carbon chains. Lambremont, E. has demonstrated the reduction of palmitic and stearic acid to the corresponding fatty alcohols and the conversion of these fatty alcohols to fatty acids in the tobacco hornworm. It is likely that a similar reduction of very long-chain fatty acids to fatty alcohols is involved in the biosynthesis of cuticular primary alcohols in insects. Similarly, one can envisage the decarboxylation of very long-chain acids to *n*-alkanes one carbon unit shorter. Since very long-chain fatty acids are not commonly found in insect cuticular lipids, it appears that they must be efficiently decarboxylated to alkanes or reduced to the corresponding fatty alcohol. Lack of substantial amounts of very long-chain n- or branched fatty acids appears characteristic of most insect cuticular lipids.

Over one-half of the secondary alcohols in the grasshopper *Melanoplus sanguinipes* would have to arise from odd-chain fatty acids for a direct condensation mechanism to be operative in their biosynthesis. Tricosan-11-ol is the major isomer comprising over a third of the secondary alcohols in this insect. The fatty acids 11 : 0 and 13: 0 would be required for a direct condensation pathway of biosynthesis to occur. Odd-chain fatty acids have not been extensively studied in insects, but unless insects possess an unusually active odd-chain fatty acid synthesizing system, the structures of the secondary alcohols make it appear unlikely that they arise directly from a condensation pathway.

Several lines of evidence from radioisotope tracer studies also argue against a condensationreduction pathway for hydrocarbon biosynthesis in insects. In the migratory grasshopper, M. s*anguinipes,* the proposed intermediates of the condensation pathway, symmetrical ketones and secondary alcohols of 23, 27 and 31 carbons, were not converted to hydrocarbons. Furthermore, the unidirectional metabolism of alkanes to secondary alcohols in this insect does not support the condensation pathway for hydrocarbon biosynthesis. Similar studies in the cockroach, *Periplaneta fuliginosa,* showed that a symmetrical C_{23} secondary alcohol and ketone were not reduced to n-tricosane, the major *n*alkane in this insect. Since the possibility of a non-symmetrical condensation of a C_{10} with a C_{14} or a C_8 with a C_{16} fatty acid was conceivable for the biosynthesis of *n*tricosane, the incorporation of 7-, 8-,10-,11- and 12tricosanols into hydrocarbon was also studied. None of these possible intermediates were incorporated into hydrocarbon in *P. fuliginosa.*

Decarboxylation of long-chain fatty acids to n-alkanes

Direct evidence in favor of the elongationdecarboxylation pathway in insects was obtained by demonstrating the direct decarboxylation of longchain fatty acids to n-alkanes in several cockroaches and in the termite *Zootermopsis angusticollis*. *In vivo* studies showed that [R-^{3}H]hexacosanoic *acid was converted to n*pentacosane in the American cockroach, and that [R-^{3}H]tetracosanoic acid was converted to *n*tricosane in *Periplaneta fuliginosa.* The decarboxylation of high specific activity [15,16-^{3}H]tetracosanoic acid in the termite Z. *angusticollis* was demonstrated both *in vivo* and *in vitro.* Whereas [1^{14}C]acetate labeled all the cuticular hydrocarbon components, [R-^{3}H]tetracosanoic acid labeled n-tricosane exclusively both *in vivo* and *in vitro*. α-Hydroxytetracosanoic acid was incorporated into hydrocarbon more efficiently than was tetracosanoic acid, suggesting that the first step in the reductive decarboxylation involves α-hydroxylation. Most of the decarboxylase activity was located in the microsomal fraction, and was stimulated two-fold by the addition of ascorbic acid.

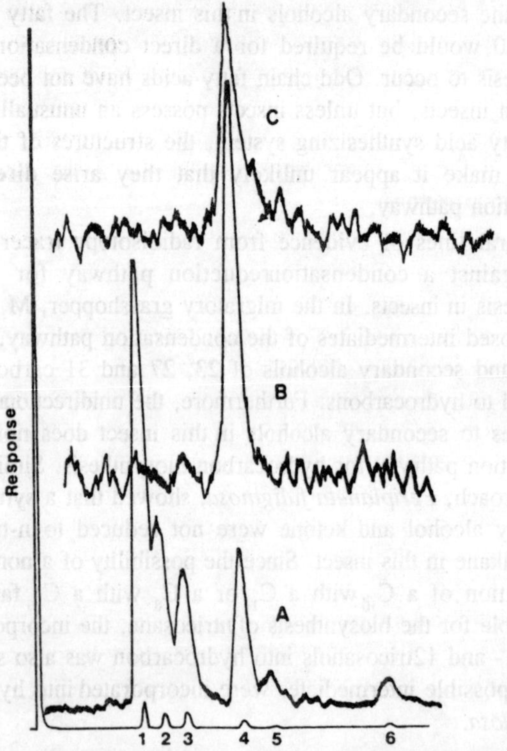

Fig. 5.5. Radio-GLC traces of the distribution of [1-^{14}C]acetate in vivo (A) and [15,16-^{3}H]tetracosanoic acid invivo (B) and in vitro (C) into the major cuticular hydrocarbons of Zootermopsis angusticollis. The bottom trace is a mass trace. Components are identified as (1) n-heneicosane, (2) 5-methylheneicosane, (3) 5,17-dimethylheneicosane, (4) n-tricosane, (5) 5-methyltricosane, and (6) n-pentacosane.

Although all the evidence to date favors an elongation-decarboxylation pathway, alternative pathways cannot be ruled out. There exists no direct evidence demonstrating the elongation of 16 or 18 carbon fatty acids to the putative precursors of hydrocarbons, and very long-chain fatty acids and methyl branched fatty acids are not usually reported in insects. This suggests that if the elongation-decarboxylation pathway does occur, the elongation and decarboxylation reactions must be tightly coupled, as significant amounts of intermediates are not released.

The decarboxylation of long-chain acids cannot be demonstrated in all insects. For example, the female housefly, which synthesizes n-tricosane, did not decarboxylate [15,16-^{3}H] tetracosanoic acid *in vivo*. Rather, most of the radioactivity from this acid was recovered in 16 and 18 carbon fatty acids after a 2 h incubation, suggesting a very efficient chain shortening mechanism (Dillwith and Blomquist, unpublished observations).

Biosynthesis of alkenes

A comparison of the data on the incorporation of [1-^{14}C]acetate, [1-^{14}C]stearate, [1-^{14}C]oleate, and [9,10-^{3}H]oleate into hydrocarbon by the housefly, *Musca domestica,* suggests a pathway in which oleic acid is elongated and decarboxylated to form the alkenes, including the sex pheromone component, (Z)-9-tricosene.

The major cuticular hydrocarbon component of the American cockroach is (Z,Z)-6,9-heptacosadiene, comprising 70% of the total hydrocarbons. Labeled acetate was incorporated about equally into the saturated and diene hydrocarbons, whereas [1-^{14}C]linoleic acid was preferentially incorporated into the diene fraction. Recent work showed that [9,10-^{3}H]oleate was incorporated almost exclusively into the diene, with less than 6% of the radiolabel being recovered in the saturated fraction. [^{13}C]NMR experiments demonstrated that [2-^{13}C]acetate labeled carbons 25 and 27 but not carbon 3 of the C_{27} alkadiene. In addition, ozonolysis of the diene labeled from [1-^{14}C]acetate followed by radio-GLC analysis showed that carbons 1-6 were not labeled. These data indicate that linoleate from the diet or synthesized *de novo* is elongated by the addition of acetate units and is then decarboxylated.

Methyl branched alkanes

Methyl branched alkanes often comprise a significant portion of cuticular hydrocarbon mixtures and have been shown to serve as both pheromones and kairomones in many insects. Consequently, the origin of the methyl branch units has received considerable attention.

2-Methylalkanes. The biosynthesis of the 2-methylalkanes has been investigated in the ground cricket, *Nemobius fasciatus,* and in the field cricket, *Gryllus pennsylvanicus*. The data suggest that the carbon skeleton of valine and leucine serve as the precursors to the even and oddnumbered carbon chain 2-methylalkanes, respectively. 2-Methylalkanes of 23 and 25 carbons comprise 20% of the cuticular hydrocarbons of N. *fasciatus* with the other 80% n-alkanes. Of the labeled acetate that was incorporated into hydrocarbon, 28% was recovered in the 2-methylalkanes and the remainder in the nalkanes. In contrast, valine was preferentially incorporated into 2-methylalkanes, with 98% of the radio-activity that was incorporated into hydrocarbon recovered in the 2-methylalkane fraction. Likewise, isobutyric acid preferentially labeled the 2-methylalkanes. This observation suggests that the cricket converts valine to isobutyric acid which is then incorporated into the even-chain 2-methylalkanes during the initial stages of chain elongation. Both odd- and even-chain 2-methylalkanes are present in the cuticular hydrocarbons of G. *pennsylvanicus*. Labeled valine, leucine, isobutyric acid and isovaleric acid were preferentially incorporated into the branched alkanes, suggesting that leucine was converted to isovaleric acid which served as the precursor to the odd-chain 2-methylalkanes.

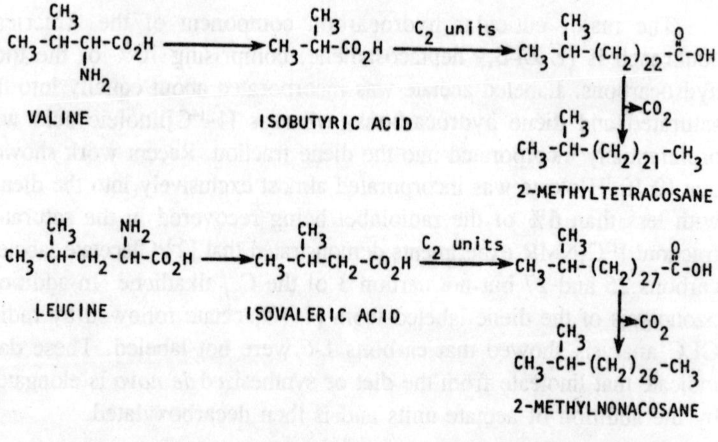

$$CH_3\text{-}CH(CH_3)\text{-}CH(NH_2)\text{-}CO_2H \longrightarrow CH_3\text{-}CH(CH_3)\text{-}CO_2H \xrightarrow{C_2\ units} CH_3\text{-}CH(CH_3)\text{-}(CH_2)_{22}\text{-}C(=O)\text{-}OH \xrightarrow{-CO_2} CH_3\text{-}CH(CH_3)\text{-}(CH_2)_{21}\text{-}CH_3$$

VALINE → ISOBUTYRIC ACID → → 2-METHYLTETRACOSANE

$$CH_3\text{-}CH(CH_3)\text{-}CH_2\text{-}CH(NH_2)\text{-}CO_2H \longrightarrow CH_3\text{-}CH(CH_3)\text{-}CH_2\text{-}CO_2H \xrightarrow{C_2\ units} CH_3\text{-}CH(CH_3)\text{-}(CH_2)_{27}\text{-}C(=O)\text{-}OH \xrightarrow{-CO_2} CH_3\text{-}CH(CH_3)\text{-}(CH_2)_{26}\text{-}CH_3$$

LEUCINE → ISOVALERIC ACID → → 2-METHYLNONACOSANE

Fig. 5.6. Proposed biosynthetic pathways for 2-methylalkanes in insects.

3-Methylalkanes. Two pathways have been suggested for the biosynthesis of 3-methylalkanes. In plants and microorganisms isoleucine is converted to 2-methylbutyric acid which is then incorporated into 3-methylalkanes during the initial stages of chain synthesis. In contrast, considerable evidence has accumulated that in several cockroaches a methylmalonyl derivative serves as the precursor to the branching methyl group of 3-methylalkanes. Isoleucine was not efficiently incorporated into 3methylpentacosane in *Periplaneta americana*.

A remarkable specificity occurs in the biosynthesis of the branched alkanes of *P. americana*, in that 3-methylpentacosane is the only

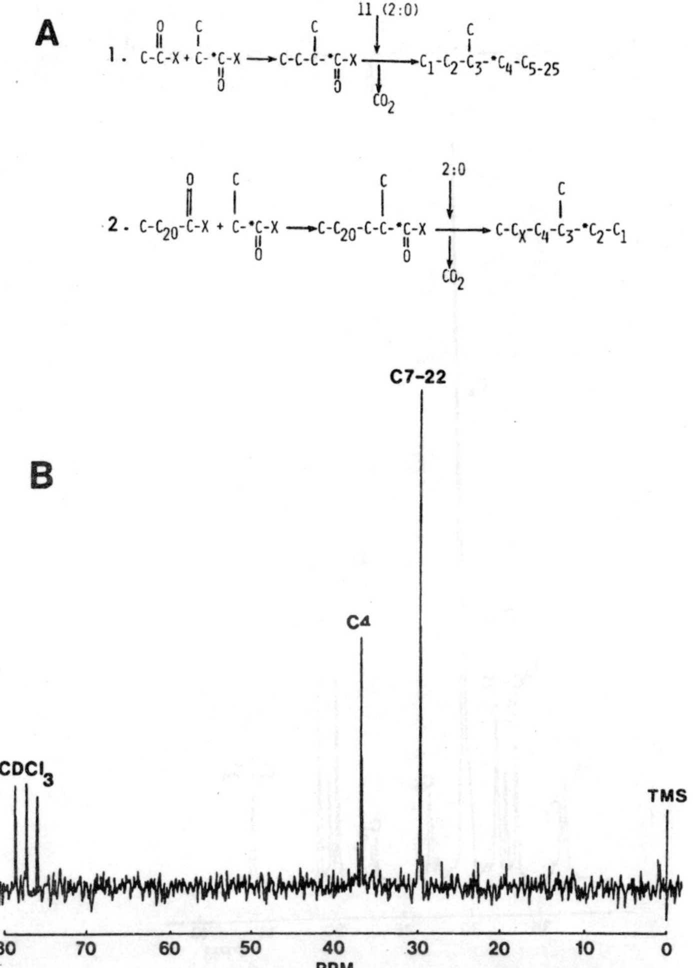

Fig. 5.7. A: Incorporation of [1-^{13}C]propionate into 3-methylpentacosane in Periplaneta americana *in early (A1) or late (A2) stages of chain elongation; B: [^{13}C]NMR spectrum of 3-methylpentacosane enriched from [1-^{13}C]propionate. Only carbon-4 is enriched, indicating that pathway (Al) in which propionate is incorporated as the second unit is operative.*

component present. The lack of methyl branched fatty acids in this insect had led to the suggestion that the methylmalonyl-CoA branching unit was added as the penultimate unit. [^{13}C]NMR experiments were used to determine if the branching unit was added as the second unit or near the end of the biosynthetic process. A knowledge of the position of the labeled carbon from [1[13]C]propionate in 3-methylpentacosane

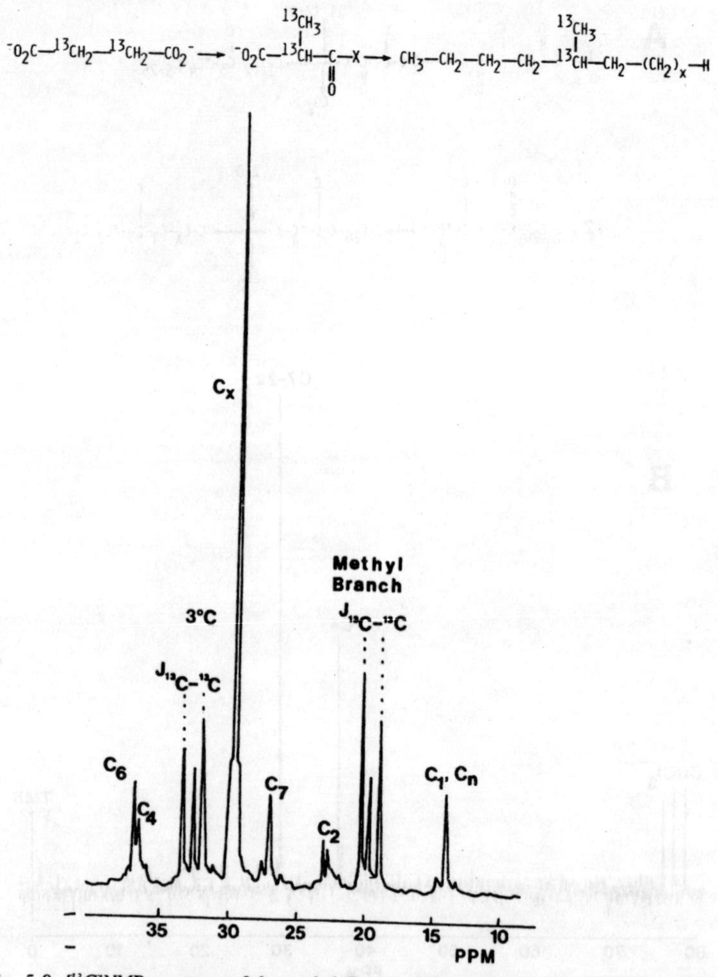

Fig. 5.8. [^{13}C]NMR spectrum of the methylalkanes of Zootermopsis angusticollis *enriched from [2,3-^{13}C]succinate. The carbon-carbon coupling shows that the labeled carbons are incorporated intact into the branching methyl carbon and the tertiary carbon.*

would allow a determination of whether propionate was incorporated in early or late stages of the elongation process. If carbon 2 was enriched, the methyl group would have been incorporated during the penultimate step of chain elongation. If carbon 4 was enriched, the methyl group would have been inserted as the second unit in chain elongation. The data from [^{13}C]NMR experiments showed that the labeled carbon from [1-^{13}C]propionate was incorporated exclusively in the 4 position, demonstrating that the methyl branch unit is added as the second unit in the elongation process. *Internally branched alkanes.* Methyl groups on the internal carbon atoms of alkane chains could originate either during the elongation process by substitution of a methylmalonyl-CoA in place of a malonyl-CoA, or by the methylation of a preformed chain. In algae, the methylation of a preformed chain by the methyl group of S-adenosylmethionine is involved in the biosynthesis of 7- and 8methylheptadecanes. The variety of positions substituted (5, 7, 9, 11, 13 or 15) and the occurrence of dimethylalkanes with methyl branches separated by 1, 3, 7, 9 or 11 methylene groups favors the methylmalonyl-CoA hypothesis in insects.

Studies on the biosynthesis of 13-methylpentacosane in *Periplaneta fuliginosa* showed that [1-^{14}C]propionate preferentially labeled this methyl branched alkane. Similarly, [1-^{14}C]propionate was incorporated almost exclusively into the mono- and dimethylalkanes in the termite *Zooterinopsis angusticollis* and the housefly, *Musca domestica.* [Methyl' ^{4}C]Methionine did not serve as an efficient precursor to the branched alkanes in any of these insects. These data suggest that propionate, as a methylmalonyl derivative, is incorporated in place of malonyl-CoA during chain elongation.

Although propionate is readily incorporated into the methylalkanes in a variety of insects, it may not be the endogenous precursor to the methylmalonyl-CoA used in branched hydrocarbon synthesis. Studies on the biosynthesis of branched alkanes in the termite, Z. *angusticollis,* suggested that succinate is a precursor to the methylmalonyl derivative which is the methyl branched precursor.

Definitive evidence that succinate is a source of the branching methyl unit of methylalkanes was obtained by using carbon-13 NMR to examine the incorporation of [2,3-^{13}C]succinate into methylalkanes in Z. *angusticollis*. The data showed that carbons 2 and 3 of succinate became the methyl branch and tertiary carbon in branched alkanes. Furthermore, the carbon-carbon coupling of adjacent labeled carbons showed that the bond between carbons 2 and 3 of succinate remains intact.

An alternative source of methylmalonyl-CoA has been demonstrated in the housefly. In this insect there was no detectable incorporation of the label from [2,3-^{13}C]succinate into the branching methyl group or tertiary carbon. In contrast, the labeled carbons from D,L[3,4,5-$^{13}C_3$]valine were incorporated intact (as determined by $^{13}C-^{13}C$ coupling) into the branching methyl carbon, tertiary carbon, and carbon adjacent to the tertiary carbon. This suggests that valine, and perhaps other amino acids that are metabolized via methylmalonylCoA, serve as a precursor to the methyl branch unit (Dillwith, Nelson, Pomonis, Nelson and Blomquist, unpublished observations).

Sodium [1-^{13}C]propionate was readily and specifically incorporated into the methylalkanes of the housefly, with an approximate 50-fold enrichment in the carbon adjacent to the tertiary carbon observed (Dillwith, Nelson, Pomonis, Nelson and Blomquist, unpublished observations). Mass spectral analysis of the [^{13}C]-enriched alkanes showed that for most components the branching unit was inserted toward the beginning of the elongation process rather than toward the end. Figure 11 summarizes our current understanding of methylalkane synthesis in the housefly, and is likely applicable to other insects.

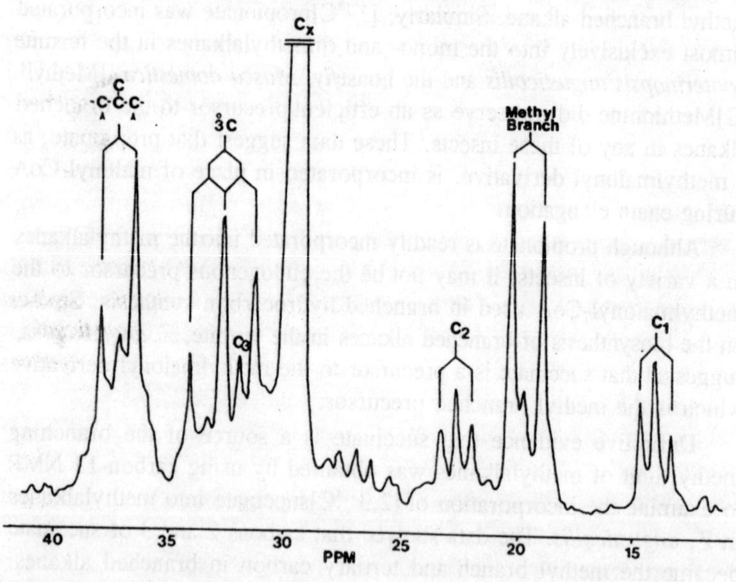

Fig. 5.9. [^{13}C]NMR of the methylalkanes of the female housefly, Musca domestica, enriched from D,1.-[3,4,5-$^{13}C_3$]valine. The carbon-carbon coupling shows that the three labeled carbons are incorporated intact into the methyl branching carbon, the tertiary carbon, and carbon adjacent to the tertiary carbon.

The choice of precursors to methylmalonyl-CoA may be related to diet. The low amounts of protein in a termite's diet makes succinate, which can be derived from carbohydrate, a logical choice. In contrast, the housefly has a much higher intake of protein, and apparently utilizes selected amino acids to form methylmalonyl-CoA. The precursor to the methyl branching group in most insects is not known.

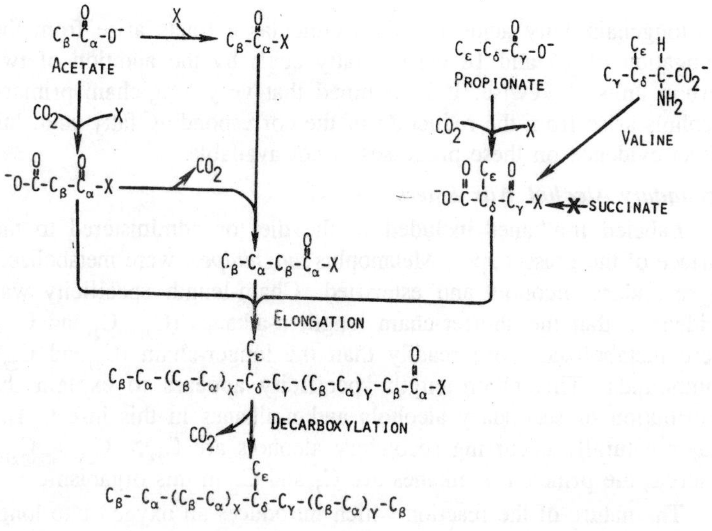

Fig. 5.10. Proposed biosynthetic pathway of the methylalkanes of the female housefly, Musca domestica. The origin of each carbon has been determined by [^{13}C]NMR spectroscopy.

In addition to being incorporated as the methyl branching carbon, the labeled carbons from [3-^{13}C]propionate and [methyl-^{13}C]methyl malonate also labeled the even-numbered carbons in the alkanes and alkenes. Likewise, [2-^{13}C] propionate labeled the odd-numbered carbons of both the alkanes and alkenes. These data suggest that propionate is converted to an acetyl derivative, with carbon 3 of propionate converted to the carboxyl carbon of acetate, and carbon 2 of propionate converted to the methyl carbon of acetate. The occurrence of this pathway for propionate metabolism in the housefly, which does not utilize the methylmalonyl-CoA mutase reaction, prompted an examination of the housefly and other insects for vitamin B_{12} levels. The housefly contained undetectable levels of vitamin B_{12} (Wakayama, Dillwith and Blomquist, unpublished observations), which was consistent with its inability to use succinate as a precursor to methylmalonyl-CoA and the presence of enzymes which directly convert propionate to an acetate derivative. In contrast, termite Z. *angusticollis* contained extremely high levels of vitamin B_{12}, which was consistent with its ability to convert succinate to methylmalonate.

Biosynthesis of Non-Hydrocarbon Cuticular Components

The biosynthesis of the non-hydrocarbon components of insect cuticular wax has not been extensively studied. It is postulated that the longchain fatty acids found in cuticular extracts arise from the elongation of 16 and 18 carbon fatty acids by the addition of two carbon units. Likewise, it is assumed that very long-chain primary alcohols arise from the reduction of the corresponding fatty acid, but direct evidence on these processes is not available.

Secondary Alcohol Wax Esters

Labeled n-alkanes included in the diet or administered to the surface of the grasshopper, Melanoplus *sanguinipes,* were metabolized to secondary alcohols and esterified. Chain-length specificity was evident in that the shorter-chain length n-alkanes (C_{21}, C_{23} and C_{25}) were metabolized more readily than the longer-chain (C_{27} and C_{29}) compounds. This chain-length specificity appears to explain the distribution of secondary alcohols and n-alkanes in this insect. The major naturally occurring secondary alcohols are $C_{23} > C_{25} > C_{21}$, whereas the principal n-alkanes are C_{27} and C_{29} in this organism.

The nature of the reaction which introduces an oxygen into long-chain alkanes was investigated in *M. sanguinipes* with $^{18}O_2$ and $H_2{}^{18}O$. Two possible mechanisms had been suggested for this reaction. A dehydrogenation followed by hydration could occur, or alternatively, a mixed-function oxidase type enzyme could incorporate one atom of oxygen from O_2. Incubation of *M. sanguinipes* with $^{18}O_2$ or $H_2{}^{18}O$ followed by mass spectrometry of the isolated secondary alcohol showed that molecular oxygen serves as the oxygen donor. The ^{18}O from $^{18}O_2$ was incorporated into secondary alcohols. This observation suggests that a mixed-function oxidase type enzyme is involved in secondary alcohol biosynthesis.

The esterification of labeled secondary alcohols occurs readily in *M. sanguinipes,* with a fairly evident chain-length specificity. The shorter chain secondary alcohols C_{21}, C_{23}, C_{25} and C_{27} were esterified more readily than the C_{31} compound. The esterification reaction was very rapid and efficient. Greater than 80% of exogenous C_{23} secondary alcohol was esterified in 18 h, and almost all the administered *n*alkane which was hydroxylated appeared in the form of a secondary alcohol wax ester. In accord with this observation, only trace amounts of free secondary alcohol were present in the cuticular extract of this insect.

Primary Alcohol Wax Esters

Labeled acetate, palmitate and tetracosanol injected beneath the cuticle of the honeybee were incorporated into wax monoester. Acetate and palmitate labeled both the alcohol and acid moieties of the monoester, whereas tetracosanol was incorporated almost exclusively into the alcohol moiety.

The esterification of primary alcohols was studied with a microsomal preparation obtained from the abdomens of insects not actively producing comb wax. Microsomal preparations (105,000 g pellet) gave about a 2-fold higher specific activity than did a 10,000g pellet or a 105,000g supernatant. When a microsomal preparation was incubated in the presence of labeled tetracosanol and palmitoyl-CoA, the distribution of products showed that most of the primary alcohol incorporated into the esters was recovered in the monoester fraction, with lesser amounts in the di- and triester fractions.

The rate of wax monoester synthesis from [$1-^{14}C$]palmitate was stimulated by CoA, ATP and $MgC1_2$ and the addition of palmitoyl-CoA resulted in a 5-fold increase in monoester synthesis from labeled tetracosanol, indicating that the acyl donor group is activated in the form of an acyl-CoA. Palmitoyl-CoA stimulated monoester formation from labeled tetracosanol and increased amounts of wax ester were formed with increasing amounts of palmitoyl-CoA up to 10-'M, after which it inhibited, presumably due to a detergent effect.

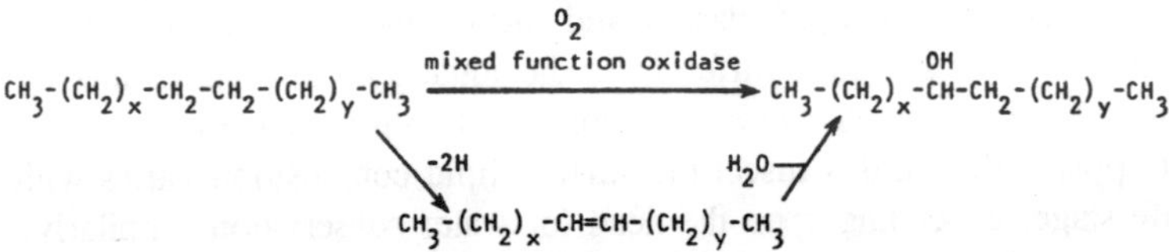

Fig. 5.11. Proposed pathway for the biosynthesis of secondary alcohols in Melanoplus sanguinipes. *Stable isotope studies have demonstrated that the oxygen in the hydroxyl group arises from molecular oxygen.*

The chain-length specificity of the primary alcohols incorporated into monoester was examined, and the results showed that labeled alcohols of 16, 18, 24 and 28 carbons were readily incorporated into monoester with the shorter-chain 16 and 18 carbon alcohols incorporated at higher rates. The naturally occurring primary alcohols are of chain lengths C_{24}—C_{32}. This suggests that the chain-length specificity of the alcohols present in monoester does not reside in the esterification step, but rather in the reduction of fatty acids to alcohols.

The synthesis of wax esters in honeybees actively producing comb wax was examined by Lambremont, E. and Wykle, R.. The postmitochondrial supernatant fraction readily esterified tetracosanol to palmitoyl-CoA and the characterization of this process suggests that the synthesis of wax esters by the specialized wax glands of the honeybee is similar to that of the epidermal tissue.

Epoxides and Ketones

The biosynthesis of (Z)-9,10-epoxytricosane and 14-tricosen-l0-one was examined in the housefly, *Musca domestica*. Both of these components, which are part of the female sex pheromone, were formed from [9,10-^{3}H](Z)-9-tricosene. Both male and female insects were able to convert (Z)-9-tricosene to the corresponding epoxide and ketone with the male possessing higher activity (Blomquist and Dilwith, unpublished data). The male housefly does not possess either of these 2 oxygenated derivatives in its cuticular lipids which is apparently due to its inability to synthesize (Z)-9tricosene.

PHYSIOLOGICAL CONSIDERATIONS

Development and Environment

The variation in insect surface lipids, as affected by developmental and environmental factors, has been studied in a few insect species. The amount and composition of cuticular lipids often changes as an insect goes through developmental stages, particularly if the insect's environment undergoes major changes. For example, the amount and the distribution of lipid classes and the composition of components within a class are quite different in the aquatic naiad form of the big stonefly, *Pteronarcys californica,* compared with the terresterial adult. It appears that in this insect the surface lipid composition varies with life stage, depending upon the need for water conservaion. Similarly, the cuticular hydrocarbons vary both quantitatively and qualitatively throughout the life cycle of the flesh-fly, *Sarcophaga bullata,* and the quantity of cuticular hydrocarbon correlates well with the water conservation needs of the insect. Marked differences were noted in the cuticular lipids of immature and adult forms of the cigarette beetle, *Lasioderma serricorne*, the black carpet beetle, *Attagenus megatoma*, two species of imported fire-ants, *Solenopsis invicta* and *Solenopsis richteri* and the desert cicada, *Diceroprocta apache.* There were marked increases in either the percentage of hydrocarbons among the cuticular lipid classes, the total amount of hydrocarbon, the composition of hydrocarbons, or all three. In the desert cicada there was approximately five times more hydrocarbon in the adult than in the nymphal exuvium, and this difference correlated with the increased transpiration potential experienced by the adult. Qualitative and quantitative differences in the cuticular lipids of different life forms of the sheep blowfly, *Lucilia cuprina* and pecan weevils, *Curculio caryae* may also be related to the need to reduce cuticular transpiration in the adult form.

The cuticular lipids of insects also vary in amount and composition depending upon the season. For example, the desert tenebrionid beetle, *Eleodes armata,* possessed more hydrocarbons and a higher percentage of long-chain components in summer than they did in winter. Similar results were obtained when winter-collected beetles were acclimated to 35°, suggesting that this change in hydrocarbon is an adaptation to an altered environment. Again, the changes appear to be related to the water conservation needs of the insect.

Diapausing pupae of the tobacco hornworm, *Manduca sexta* secrete three times as much surface wax as do non-diapausing pupae. The extra thickness of the wax layer apparently protects the insect from desiccation and the authors speculate that the deposition of additional wax may result from hormonal changes accompanying entry into diapause.

Very little is known about the general area of the control of hydrocarbon biosynthesis in insects. Work by Armold, M. and Regnier, F., however, suggests that 20-hydroxyecdysone stimulates hydrocarbon formation. They found that in the development of *S. bullata* there are two periods of rapid accumulation of hydrocarbons. The most rapid accumulation is during the 4-day period preceding the pupal-adult ecdysis (> 20 pg day-' insect-'). An earlier accumulation (5-81 cg day' insect) occurs during pupariation and the 3-day period following pupariation. At pupariation, exogenous 20-hydroxyecdysone was shown to stimulate hydrocarbon biosynthesis from [^{3}H]acetate at a rate of 1.3 times the rate of controls 10.5 h after hormone administration and at a rate of 3.3 times the rate of controls 24 h after hormone administration. In an *in vitro* study they showed that integuments of 20-hydroxyecdysone-treated insects incorporated acetate into hydrocarbons 9.8 times better than integuments of control insects. This shows that cuticular hydrocarbon biosynthesis in S. *bullata* occurs in the integument, and that the locus of regulation is also present in the integument.

Dramatic changes in the composition of the cuticular lipids of the housefly, *Musca domestica,* occur with age in the adult insect. Newly emerged males and females have almost identical cuticular compositions. Dramatic changes take place in the female between days 2 and 3. (Z)9-Tricosene, (Z)-9,10-epoxytricosane, and (Z)-14tricosen-10-one first appear at this time, and increasing amounts of these components and methylalkanes accumulate in the female as the insect ages (Dillwith, Nelson and Blomquist, unpublished observations). These components, which are part of the housefly sex pheromone, are absent, or present in very small amounts, in the male. Because of the apparent similarity in timing of the beginning of pheromone production and ovarian maturation, the production of the pheromone components was correlated to ovarian development.

Production of the cuticular sex pheromone components of the female housefly, including (Z)-9tricosene (muscalure), (Z)-9,10-epoxytricosane, (Z)-14-tricosen-l0-one, and methyl branched alkanes, correlated with ovarian development. No pheromone was produced by females with ovaries in pre-vitellogenesis, production was initiated in imsects during early vitellogenesis, and larger amounts of pheromone were found in females as the ovarian follicles matured (Dillwith, Adams and Blomquist, unpublished observations).

Females ovariectomized shortly after adult emergence did not produce pheromone. Reimplanting pre-vitellogenic ovaries into ovariectomized females and allowing them to reach early vitellogenic stages restored pheromone production. Implanting ovaries into male insects resulted in altering the composition of their cuticular lipids such that they then produced (Z)-9-tricosene (muscalure), the C_{23} epoxide and ketone, and large amounts of methylalkanes (Blomquist, Adams and Dillwith, unpublished data). Injection of 20-hydroxyecdysoneinto ovariectomized females or into newly emerged males caused the same alteration in cuticular lipids as did implanting of ovaries, and production of pheromone components was initiated. The major cuticular component produced by ovariectomized females and control males is (Z)-9-heptacosene. The presence of maturing ovaries or injection of 20hydroxyecdysone causes a change in the chain length of the alkenes such that (Z)-9-tricosene is produced, and increases the amount of methylalkanes synthesized. Adult male and female insects both possess the enzyme(s) which convert (Z)-9-tricosene to the corresponding epoxide and ketone at all ages, and the production of these oxygenated derivatives occurs when substrate ((Z)-9-tricosene) is present (Blomquist, Adams and Dillwith, unpublished data).

Biosynthesis of Beeswax: Effect of Season, Age

In addition to producing a thin layer of surface lipids, some insects, such as the honeybee, produce a much larger amount of wax for the honeycomb. The biosynthesis of beeswax has been investigated by Piek, T. and Young, R., who showed that labeled acetate was incorporated into the hydrocarbon, ester and acid fractions of bees- wax. The biosynthesis of wax by the honeybee was studied in winter and summer. It was observed that in winter, bees not only incorporated acetate into hexane-extractable lipids at much lower levels than in summer, but

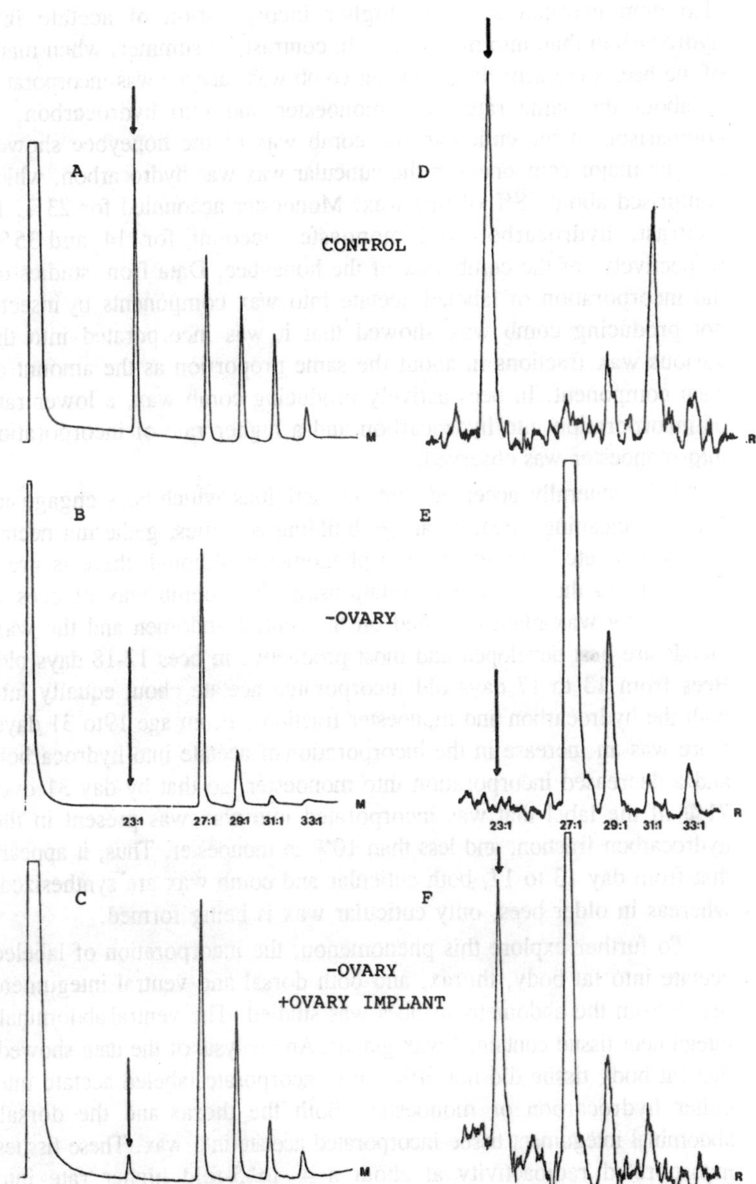

Fig. 5.12. The effect of ovariectomy and re-implanting ovaries on alkene synthesis in the female housefly, Musca domestica.

also demonstrated a 4-fold higher incorporation of acetate into hydrocarbon than into monoester. In contrast, in summer, when many of the bees were actively producing comb wax, acetate was incorporated at about the same rates into monoester and into hydrocarbon. A comparison of the cuticular and comb wax of the honeybee showed that the major component in the cuticular wax was hydrocarbon, which comprised about 58% of this wax. Monoester accounted for 23%. In contrast, hydrocarbon and monoester account for 14 and 35% respectively, of the comb wax of the honeybee. Data from studies on the incorporation of labeled acetate into wax components by insects not producing comb wax showed that it was incorporated into the various wax fractions in about the same proportion as the amount of each component. In bees actively producing comb wax, a lower rate of incorporation into hydrocarbon and a higher rate of incorporation into monoester was observed.

It is generally accepted that the activities which bees engage in (i.e. cell cleaning, feeding larvae, building activities, gathering nectar and pollen, etc.) are age-related phenomena, although there is great flexibility in the age-activity relationship. The comb wax of bees is secreted by wax glands located on the ventral abdomen and the wax glands are best developed and most productive in bees 12-18 days old. Bees from 13 to 17 days old incorporated acetate about equally into both the hydrocarbon and monoester fractions. From age 19to 31 days there was an increase in the incorporation of acetate into hydrocarbon and a decreased incorporation into monoester, so that by day 31 over 71% of the label that was incorporated into wax was present in the hydrocarbon fraction, and less than 10% in monoester. Thus, it appears that from day 13 to 17, both cuticular and comb wax are synthesized, whereas in older bees, only cuticular wax is being formed.

To further explore this phenomenon, the incorporation of labeled acetate into fat body, thorax, and both dorsal and ventral integument tissue from the abdomens of bees was studied. The ventral abdominal integument tissue contained wax glands. An analysis of the data showed that fat body tissue did not efficiently incorporate labeled acetate into either hydrocarbon or monoester. Both the thorax and the dorsal abdominal integument tissue incorporated acetate into wax. These tissues incorporated radioactivity at about a 4- to 5-fold higher rate into hydrocarbon than into monoester. In contrast, the ventral abdominal tissue, which contained the wax glands, incorporated radioactivity about equally into both hydrocarbon and monoester. These data demonstrate that the site of wax synthesis in the honeybee is associated with the integument and points out the difference in the wax produced by the glands (comb wax) compared with the cuticular wax.

CONCLUDING REMARKS

The development of microanalytical techniques, particularly GLC-MS, has allowed rapid advances in our understanding of the chemistry of insect cuticular lipids. However, our knowledge of the chemical composition of insect cuticular lipids is still limited to representative species of only seven orders, and in many cases is restricted to one lipid class—the hydrocarbons. Although it has been known for decades that a major function of insect cuticular lipids is to restrict water loss, a generally accepted model for the observed waterproofing is not available. Indeed, the recent data which argue against the long-held view of an "oriented monolayer" should result in a renewed interest in gaining an understanding of the waterproofing properties of cuticular lipids. The recognition that cuticular lipid components function in chemical communication in certain species has resulted in a great deal of interest to determine how general this phenomenon is, and to examine the types of roles cuticular lipids play. The general metabolic pathways have been determined for some of the cuticular lipid components, and information is becoming available on the endocrine regulation of cuticular lipid formation in a few species. However, these studies have only begun to explore the processes by which cuticular lipid components are formed and regulated and many intriguing questions remain. A strong case can be made for the joining of forces between behavioral ecologists, physiologists, morphologists, biochemists and chemists to better our understanding of insect cuticular lipids.

FAT BODY

In biology it is axiomatic that structure reflects function. Fat body is so pleiomorphic and physiologically versatile a tissue that it would be impossible to describe its structure in the singular as our title suggests. Indeed, Wyatt, G. has stated that the functional diversity of fat body cells may be unequalled by any other metazoan cell type. An understanding of the morphology of the fat body is therefore not accomplished by a simple description of structure, as it changes during development. The appropriate physiology has yet to be discovered for much of the structure that has already been described. The fat body varies in structure between organisms, between stages, according to its internal location and by differentiation from cell to cell. It changes its structure cyclically in relation to growth in the intermoult/ moult cycle, in relation to changes in commitment at metamorphosis, in adaptation to stress, nutrition and metabolic demand and in relation to reproduction. In this account we point out how the predictable cyclical activities make the fat body useful for studying several problems of cell biology in addition to those relating directly to insect physiology.

Glossary of Cell Biology Terms

The fat body is increasingly being recognized as a tissue suitable for studying aspects of cell biology that are difficult to investigate with vertebrate cells. Fat body development is cyclical and sequential compared to liver, which is in a dynamic equilibrium. The fat body is thus convenient for observing evanescent processes such as autophagy or mitochondrial division, and it is the only material that has been used to study the Golgi complex beads.

Autophagic vacuole (A V) A 2° lysosome derived from isolation bodies and 1° lysosomes. Synonyms: cytolysome, cytosegresome.

Confronting cisternae RER apposed to the plasma membrane or to peroxisomes at localized regions free of ribosomes.

Cortex Cytoplasm below the plasma membrane containing microfilaments and other cytoskeletal elements to the exclusion of ER and ribosomes.

Dense body Descriptive term for many inclusions that have been insufficiently studied for satisfactory characterization.

Endoplasmic reticulum (ER) Membrane-bound compartments continuous with the nuclear envelope that connect with Golgi complexes (GC) through transition vesicles (tv). All fat body ER bears ribosomes (RER). Smooth ER is abundant in oenocytes but absent from fat body which differs from liver in this respect.

Golgi complex (GC) Fat body GCs consist of rings of GC beads on the RER where transition vesicles arise, three or four saccules together with secretory vesicles (SVs) and 1° lysosomes at the secretory face.

Golgi complex beads (GCb) 10 nm particles arranged in rings around the transition vesicles on the smooth face of RER where GCs arise.

Heterophagic vacuole Membrane-bound vacuole containing extracellular material scheduled for lysis. Derived from endocytosed material and 1° lysosomes, i.e. a 2° lysosome.

Instar A developmental stage characterized by an integument outwardly unchanged between ecdyses, but with a fat body that varies with age in the stadium.

Intermoult/moult cycle The developmental sequence which repeats itself from stadium to stadium. In the fat body it begins with cell preparation for intermoult syntheses (such as perioxisome formation, RER and GC formation, mitochondrial growth and division), is followed by intermoult syntheses (such as lipid, glycogen,haemolymph protein), and concludes with ecdysial events (such as tyrosine mobilization) and preparations for the next stadium.

Isolation body (IB) The fragment of cytoplasm isolated by the paired isolation membranes. It becomes an autophagic vacuole after fusing with a 1° lysosome. Several IBs often fuse together.

Isolation membranes The paired membranes of the envelope which invests organelles scheduled for lysis.

Lymph Fluid in intercellular spaces below the basal lamina through which it has filtered from the haemolymph.

Multivesicular body (MVB) A structure bound by a single unit membrane containing material carried to it by pinocytosis vesicles and 1° lysosomes. MVBs have inner vesicles arising by invagination at their surface and are concerned in membrane turnover and protein digestion.

Peroxisome A particle with a single unit membrane enclosing dense contents and a core, and which always contains catalase and oxidases. Synonym: microbody.

Pinocytic vesicle Microvesicle arising at the plasma membrane surface that is involved in conveying membrane and protein into the cell.

Plasma membrane reticular system (PMRS) The system of plasma membrane infolds forming an extra inner surface in many large insect cells.

Primary lysosome (1° L) Microvesicles from GC carrying lytic enzymes or proenzymes.

Protein granule (PG) Membrane-bound vesicle storing protein that has been pinocytosed from the haemolymph. The protein is often crystalline. PGs are often composite structures, having fused with AVs.

Provacuoles Plasma membrane derived compartments that fuse together to form tyrosine storage vacuoles.

Residual body Membrane-bound inclusion characterized by undigested residues (membrane fragments or whorls, myelin figures, ferritin-like particles, etc.)

Secondary lysosome (2° L) Membrane-bound vesicle resulting from fusion of primary lysosome with vacuole carrying material to be digested.

Secretory vesicle (SV) Membrane-bound vesicle of protein destined for secretion outside the cell by exocytosis. Synonyms: secretory granules, zymogen granules.

Stadium The period from one ecdysis to the next, during which the fat body changes in the characteristic intermoult/moult sequence of development.

Transition vesicle (tv) Microvesicle between the RER and the outer saccule of the Golgi complex. They bud off through the centre of the GCb rings.

Tyrosine vacuole Large vacuole storing tyrosine in larval fat body cells.

Urate granule Membrane-bound structure at metamorphosis storing urate and uric acid sometimes with fibrous and vesicular contents. Of uncertain origin.

Urate vacuole Large vacuole storing urate in cockroach urate cells. Structurally similar to large urate granules in Lepidoptera.

Vacuole Large membrane-bound structure which in the fat body may store either tyrosine or urate. Synonym: watery vacuole.

Vitellogenesis Synthesis of vitellogenin by fat body.

Abbreviations

AV	Autophagic vacuole
B	Bacteria
BPM	Bacterial plasma membrane
BL	Basal lamina
cv	Coated vesicle
D	Desmosomes
E	Envelope of Gram-negative bacterium
G	Glycogen
S	GC saccule
GC	Golgi complex
GCb	Golgi complex beads
H	Haemocoel
HP	Haemolymph proteins
HD	Hemidesmosome
L	Lipid droplet
LP	Lymph proteins
Ls	Lymph space
mf	Microfilament
mt	Microtubule
M	Mitochondria
MVB	Multivesicular bodies
N	Nucleus
P	Peroxisomes
PMRS	Plasma membrane reticular system
PM	Plasma membrane
PG	Protein storage granules

PV	Provacuole
RER	Rough endoplasmic reticulum
SV	Secretory vesicle
tv	transition vesicle
UG	Urate storage granules
UV	Urate vacuole
C	Urate vacuole core
V	Vacuole
VM	Vacuole membrane
1° L	1° lysosomes

Preservation of the Fat Body

The. fat body is an ungrateful kind of cell for structural studies as Dutkowski, A. pointed out. Published electron micrographs attest to the difficulty of fixation. There are many reasons for this. Fat body usually contains different kinds of vacuoles and compartments that require different conditions for optimal fixation of the natural shape. It may contain phagic vacuoles, that if broken, liberate hydrolytic enzymes into the cell. It nearly always contains large areas of lipid and glycogen that lack the mechanical stability given to regions of cytoplasm containing a cytoskeleton. Lastly, it is exposed in the haemocoel, its natural shape being partly due to tension on the basal lamina strands that connect it to other tissues. The conditions appropriate for good fixation are similar to those needed for the epidermis. The following details are often neglected:

(1) Fixation should be by inflation, preferably after some haemolymph has been removed to allow replacement without mechanical distortion. Fat body should not be dissected in saline prior to fixation.

(2) The initial fixation should be in a small volume, since large volumes leach out cell components. This is best accomplished by leaving an injected insect for several minutes before opening it up.

(3) The pH should be kept on the alkaline side of 7.4 since fixation is much slower close to neutrality.

(4) Fixation and other steps prior to in-block staining should be on ice, since cold discourages the solution of components before they can be fixed.

(5) The fat body should be carried through to embedding with its natural shape protected within a ring of integument.

Good preservation is particularly important for distinguishing between the partial autolysis that is part of metamorphic cell reprogramming and cell death preceding replacement by stem cells. Poorly preserved cells undergoing organelle autophagy may sometimes have been misinterpreted as dying larval cells.

STRUCTURE

The Diversity of Tissue Arrangement and Gross Morphology in Different Orders

The fat body is typically located in two body regions that reflect its embryonic origin: close to or surrounding the gut, and adjacent to the integument. In most orders, segmentally arranged somites with a central coelomic cavity appear in the embryo. As the coelomic cavity becomes continuous with the haemocoel the splanchnic mesoderm of the somite grows around the gut and some cells differentiate to form perivisceral fat body. The somatic mesoderm of the somite contributes cells which form the peripheral fat body next to the integument. These layers are not always discrete: continuity between perivisceral and peripheral fat body is frequently encountered and the initial segmental arrangement is typically lost in later development.

The structure and arrangement of fat body is characteristically constant within a species, and members of the same order usually show the same general plan. Descriptions of fat body arrangements and gross morphology. Whatever the precise layout of the tissue, a constant feature is that a very large area is exposed to the haemolymph which allows rapid metabolite exchange.

Basic Cell Structure: Trophocytes

In some orders, only one type of cell is found in the fat body: the trophocyte or adipocyte. In other orders trophocytes are most numerous but the fat body is an aggregate of two or three cell types. During embryogenesis and early postembryonic development trophocytes have a rounded nucleus which may be polyploid, and the cytoplasm contains few inclusions. Thereafter, trophocyte morphology varies with nutritional state, with the time in the intermoult/moult cycle and also with the developmental stage of the insect. At times of great metabolic activity much rough endoplasmic reticulum (RER) and numerous Golgi complexes (GC) occur together with abundant mitochondria and peroxisomes. This contrasts sharply with the start of non-feeding periods when the fat body may be laden with stores of protein, lipid and glycogen.

In some species, for example *Calpodes ethlius,* the trophocytes appear to differ from each other only in size. Peripheral cells are smaller than those in the perivisceral ribbons, which probably only reflects differences in ploidy. Their activities are similar and more or less synchronous. In other species the trophocytes show regional specialization.

TABLE 6.1: GROSS MORPHOLOGY OF FAT BODY TISSUE

Order and species	*Observations*
Coleoptera	
Leptinotarsa decemlineata	Tissue consists of a few small lobes, some next to gonads and some near integument. Peripheral lobes consist of an oenocyte surrounded by trophocytes. (A)t
Dictyoptera	
Periplaneta americana *Diploptera punctata*	Flat lobes, mycetocytes centrally placed and more or less surrounded by urate cells. Trophocytes tend to occupy edge of lobe. (A)
Diptera	
Calliphora stygia	(L) Flat sheets, one cell thick in paired anterior and dorsolateral lobes. Posterior lobe is one cell thick, highly fenestrated and folded. Cell number about 11,500 throughout larval life. Progressive histolysis from anterior to posterior. (L)
Sarcophaga arg,rostoma	Cell separation starting at anterior is completed within 48 h of puparium formation. (L)
Calliphora erythrocephala	Oenocytes among fat body trophocytes. (A)
Glossina austeni	Oenocytes among fat body trophocytes. (A)
Sarcophaga bullata	Oenocytes among fat body trophocytes. (A)
Dacus tryoni	Oenocytes among fat body trophocytes. (L,A)
Musca domestica	Oenocytes among fat body trophocytes. (A)
Ephemeroptera	

Heptagenia sp. *Chirotonetes* sp.	Thin irregular sheets broadening to form irregular pads around the spiracles. (N)
Hemiptera	
Rhodnius prolixus	Abdominal tissue is lace-like, predominantly a single cell layer but thicker at lateral margins. Spaces in tissue reduced by increasing cell number in successive instars. (N).
Lepidoptera	
Pontia rapae *Phlegethontius quinquemaculata* *Melitta satyriniformis*	Perivisceral tissue in longitudinal ribbons which may be much folded and cut. This tissue attached in places to smaller quantity of tissue adjacent to integument. (L)
Calpodes ethlius	Sheets of cells one layer thick growing to two layers thick in each stadium. (L) During metamorphosis larval tissue disaggregates into contiguous clumps of cells that reorganize into nodular clumps surrounding tracheoles or developing gonads. (A)
Orthoptera	
Schistocerca gregaria	Throughout abdomen and thorax.
Locusta migratoria *Melanoplus femur-rubrum* *Orchelimum* sp.	Peripheral trophocytes interspersed with oenocytes firmly attached to integument, central loose network of anastomozing lobes in haemocoel that is frequently continuous with peripheral tissue. (N,A)
Gryllus abbreviatus	
Trichoptera	
Hydropsyche sp.	Many leaf-like bladders, some forming separate masses while others are connected at their bases by tissue lying close to body wall. (L)

Trophocyte Shape and the Intercellular Environment

The form of the fat body as a tissue depends upon three factors: the basal lamina, junctions, and the shape of each cell determined by its cytoskeleton.

The basal lamina

Except for the haemocytes, all tissues in an insect are separated from the haemolymph by a basal lamina—including even those tissues such as fat body, epidermis, ovarioles, wing discs and pericardial cells that traffic in haemolymph proteins. Of these, the fat body stands out as a tissue that must both take up and secrete proteins through the basal lamina. The appearance of the basal lamina suggests that it may serve as a barrier to some haemolymph components such as cells and large proteins. The basal lamina around Malpighian tubules, for example, only allows the smallest particles of colloidal gold to penetrate as far as the cell surface. For this reason we propose to refer to the spaces around fat body cells and below the basal lamina as lymph spaces, since the fluid in them may be filtered from the haemolymph bathing the tissue. A property of the fat body basal lamina must be permeability or selective permeability to haemolymph proteins in both directions. There is nothing in the appearance of the basal lamina that gives a clue how this may be managed. The texture may be loose enough to allow smaller proteins to pass through, but some of the larger polymers may well be excluded. The main larval storage proteins synthesized by the fat body in *Calpodes* have molecular weights of 720, 580, and 470 K daltons or more and are polymers of six or more peptides with molecular weights of 70-80 K daltons. Negative staining shows the smallest of these large polypeptides to be 11-12 nm in diameter making it unlikely that they can traverse the basal lamina unimpeded. If these large polymers are excluded then it may be that they pass through the basal lamina as monomers with which the polymers are in equilibrium.

The basal lamina around tissues such as nerves is layered, the number of layers increasing from stadium to stadium as though they have been added on at each stage. The fat body (and other tissues that grow extensively) on the other hand, have single-layered basal laminae without extra layers in later stages. Since the basal lamina does not get thinner, material must presumably be added to it as the fat body grows. Why then is it not layered like that around nerves?

The absence of layering is probably the result of involution of the old stage basal lamina and the secretion of a new one at each moult. Immediately after the third to fourth ecdysis in *Calpodes,* the basal lamina is endocytosed in large vacuoles within which it disappears, presumably as a result of digestion. The basal lamina is held to the plasma membrane by hemidesmosomes. These concentrate in patches that become vacuoles, dragging the basal lamina into the vacuole lumen with them. This phagocytosis of the basal lamina occurs both at the outer surface and between cells.

The relationship between the basal lamina and the fat body is a complicated one, for it is both at the haemolymph face of the tissue and threaded through the interior between the cells in reticular strands. The strands are particularly obvious at the interior meeting places where three or four cells may join together. There is a continuous gradation between desmosomes joining cells together directly, paired hemidesmosomes abutting the basal lamina, unpaired hemidesmosomes of adjacent cells attached to the same basal lamina and single separate hemidesmosomes. The close association between the basal lamina and desmosomes suggests that the fat body has a direct role in basal lamina secretion. It is difficult to see how haemocytes, that may be the source of basal laminae, could be involved in joining desmosomes with basal lamina deep within the fat body tissue. No haemocyte associations with desmosomes have been observed.

The fat body is strung to other tissues by strands of basal lamina that often have a core of elastic fibres.

Junctions

Junctions between fat body cells are desmosomes and hemidesmosomes, gap junctions and occasionally rather long-lived mid bodies. The electrical coupling of *Calpodes* larval fat body cells varies with stage in the intermoult/moult cycle. Although most of the coupling is presumably due to the gap junctions, it is relatively easy to observe the mid bodies remaining after cell division during intermoult preparation. Mid bodies must therefore remain intact for many hours before cell separation is complete.

Cell shape and the cytoskeleton

Fat body cells are not simple polygons strung in strips, or one or two layered sheets as they are often depicted, but have differentiated outer surfaces and an inner relationship with basal lamina strands and lymph spaces. There are two categories of outer surface differentiation dependent upon the cytoskeleton: surface processes, and infolds that give rise to the plasma membrane reticular system (PMRS). Within the fat body there are transport routes along the intercellular surfaces leading to lymph spaces.

Surface processes. In *Calpodes* and other insects the epidermis forms basal extensions or feet whose elongation and shortening is related to changes in shape of the integument. Fat body cells in *Calpodes* and

Periplaneta develop similar surface processes. In *Calpodes* fifthstage larvae they arise at the end of the intermoult preparation after the formation of a bilayered strip of tissue, at the same time as they develop in the epidermis. They extend over neighbouring cells forming stubby interdigitations that increase the surface area between the surfaces of adjacent cells.

In folds and the PMRS. Fat body cells are like oenocytes and other giant cells in having a subsurface PMRS. Early in the fourth, fifth and imaginal stadia of *Calpodes*, *Sarcophaga*, *Calliphora*, and *Leptinotarsa* the fat body cells have little surface differentiation. The surface also lacks specialization after periods of starvation. During intermoult preparation the plasma membrane folds in from the surface creating interconnected channels. These infolds may have confronting ER cisternae. In the fourth stage *Calpodes* provacuoles arise from these infolds as well as from the external surface. The irregular infolds mature into a precise meshwork of extracellular spaces forming a subsurface connected by infolds to the surface, the plasma membrane reticular system (PMRS). The subsurface is parallel to the surface and separated from it by a constant distance of 1-1.5 *μm*. The spaces of the PMRS also have rather constant dimensions with membrane to membrane separations of 100 to 150 nm in both the subsurface and its connections. The gateways between the outside and the intercellular environment also have constant dimensions but we have been unable to see a perforate diaphragm of the kind that occurs in pericardial cells and between the podia of kidney cells. The PMRS may extend laterally around the fat body cells and even as far as the lymph spaces.

These surface differentiations can be related to development of the cytoskeleton. All over the surface directly below the plasma membrane, bundles of microfilaments and microtubules form a cortex upon which the desmosomes and hemidesmosomes can insert. Microfilaments and microtubules also form a meshwork around the PMRS. The cytoskeleton is particularly obvious around the lymph spaces, as though it is concerned with their maintenance. Filaments may also be involved. Fat body cells in both *Calpodes* and *Periplaneta* occasionally contain skeins and whorls of filaments. In tissue culture much of the surface differentiation is lost, and balls of filaments appear in the cytoplasm.

Lymph spaces

The picture of the fat body which emerges is not one of cells packed tightly into a solid tissue but of cells having a peripheral sponge structure, separated by lymph spaces, and linked to one another or to the basal lamina by (hemi)desmosomes. The lymph spaces naturally contain material resembling fixed protein. These lymph spaces are not merely accumulations of exocytosed protein synthesized by the fat body. Horseradish peroxidase injected into the haemocoel appears in the lymph spaces, which must therefore receive material from the haemolymph as well as contribute material to it. This poses the question of the way that the lymph spaces communicate with the haemocoel and what their functions may be.

Intercellular surface transport routes

In cockroach and *Calpodes* intermoult fat body there are evenly spaced surfaces between all the cells. The desmosomes serve not only to hold the cells together at particular points but also to maintain a rather constant separation between cells. The separation between cells is similar to the separation between plasma membrane faces in the PMRS and is not the standard 20 nm separation commonly seen in epithelia. Fat body cells are not bonded together but bonded apart. In many places these intersurface compartments and the nearby PMRS contain packets of fixed protein presumably en route to and from the sites of pinocytosis and exocytosis that occur even deep within the tissue. Golgi complexes, MVBs and protein storage granules are all found squeezed between the interstices of lipid droplets close to plasma membranes but with little cytoplasmic communication with the rest of the cell. The structure suggests that fat body cells both need and use the spaces between them for the transport of protein. The mechanism for protein movement through these inter-surface compartments which are only 100 nm wide for distances as great at 100 lcm can scarcely be unaided diffusion.

This description of the fat body as clusters of cells surrounded by lymph with moving packets of protein between them confirms the appropriateness of the term applied to them - trophocytes.

Cell Diversity

Chromatocytes

The integument of aquatic larvae of the Simuliidae and Thaumaleidae is transparent. Their colour pattern is determined by the distribution of pigmentladen cells lying beneath the epidermal basal lamina. These chromatocytes are flat and so thin that the nucleus causes a distinct bulge in the centre of the cell. Chromatocytes are connected in sheaths, one cell thick, which surround ribbons of trophocytes but remain separated from them by a basal lamina. The cells increase in number through successive larval instars but maintain the same relative position. At metamorphosis, changes in the insect colour pattern depend on mass migrations and redistribution of these cells. Although chromatocytes have a distinctive appearance they are probably modified trophocytes. Like trophocytes they accumulate lipids during larval growth and these reserves are depleted during adult development.

Urocytes

Although uric acid is excreted into the lumen of Malpighian tubules in insects it is also commonly found stored in internal tissues, particularly the fat body. This feature is common in cockroaches where uric acid is stored thoughout life and may comprise as much as 10% of the dry weight. Urates do not accumulate in trophocytes or mycetocytes of the cockroach fat body but in cells specialized for urate storage, the urocytes. The cytoplasm of urocytes has little ER and few mitochondria, and consists mainly of very large vacuoles containing crystalloid spherules of uric acid or urate that are birefringent under plane polarized light. Such uric acid storage in vacuoles has often been called storage excretion, but the ease with which the uric acid may be mobilized in fat body deposits suggests that it may serve as a reservoir of nitrogen. An early study by Bodenstein, D. showed that, in *Periplaneta americana,* urates disappeared from the fat body when the corpora cardiaca were removed. Corpora cardiaca replacement restored urates to the urocytes, suggesting humoral control of protein or urate metabolism. However, the situation is complex because urate is redeposited in urocytes of cardiectomized insects during starvation. Uric acid is deposited in the fat body of *Periplaneta* maintained on a high-protein diet. The deposits are absent in the roaches switched to a pure dextrin diet. In a cytological study, Cochran, D. and coworkers also showed that the quantity of stored urate is related to dietary nitrogen levels. In addition, these authors demonstrated that the urate spherule is not a homogeneous structure but contains an irregular electron-dense core. A role for this structure in urate deposition or mobilization is suggested but is significance awaits further analysis. Distinct urocytes are not restricted to the Dictyoptera; they occur also in the fat body of Collembola and Thysanura.

Urate storage vacuoles of urocytes resemble the urate storage granules of lepidopteran fat body except that they are very much larger. Both may have plasma membrane-derived vacuole membranes, that in the Lepidoptera are derived from pinocytic vesicles. The urocyte vacuole membranes have the thickness of plasma membranes rather than the ER as do the tyrosine storage vacuoles.

Mycetocytes

Many insect species possess intracellular microorganisms contained within specialized cells, the mycetocytes. It has long been assumed that this relationship is symbiotic for the following reasons. The non-pathogenic micro-organisms are present in each individual, they are required for normal development in the absence of dietary supplements, and elaborate mechanisms have evolved for transmitting them from one generation to the next. A general discussion of symbiosis lies outside the scope of this chapter but a review by Houk, E. and Griffiths, G. lists references to earlier reviews. In cockroaches, some Hemiptera such as the rice planthopper, *Laodelphax striatellus*, and the primitive termite *Mastotermes*, the mycetocytes are scattered singly throughout the fat body, forming an integral part of the tissue. In cockroaches the mycetocytes are characteristically located near the centre of the flattened fat body lobes. The cells are usually large with a multilobed nucleus but the volume of cytoplasm is reduced since it is often densely packed with symbiotes. The symbiotes occupy their own compartment surrounded by host cell membrane. Each symbiote is bounded by a plasma membrane and a complex envelope consisting of a cell wall of peptidoglycan and the structured outer layer, similar in appearance to a unit membrane, that is characteristic of Gram-negative bacteria. In *Blattella germanica,* mycetocytes differentiate even in aposymbiotic embryos; that is, in embryos developed free of symbiotes. The "empty mycetocytes" have large nuclei but only small amounts of fibrous appearing cytoplasm. These authors also reported that normal mycetocytes undergo mitosis in the second half of each larval stadium. In the absence of symbiotes the cells remain small and do not divide. If the initial symbiote population is kept artificially small, the mycetocytes may grow to unusually large sizes without cell division, even though there appears to be no inhibition of symbiote replication. Mycetocytes frequently contain areas of stored glycogen but lipid droplets are rare and protein granules absent. Wuest also observed that the quantity of glycogen seemed unrelated to phases of the reproductive cycle, in marked contrast to the fluctuations seen in trophocytes.

Although mycetocytes may be derived from fat body cells, in other insects they are located in specialized organs called mycetomes that are separate from the fat body. They may be connected with the gut, gonads or fat body. An attempt has been made to show a symbiotic relationship between host and microorganism in the pea aphid, *Acyrthosiphon pisum,* where the mycetome is a distinct abdominal organ lying dorsal to the gut among the fat body lobes. Large central cells of the mycetome containing one kind of microorganism are partially surrounded by flattened sheath cells containing a second type. Electron microscope autoradiography coupled with the incorporation of [^{3}H]mevalonate, suggested that both types of micro-organism synthesize cholesterol. The majority of silver grains were at first associated with the symbiote membranes but the frequency of grains over surrounding mycetocyte tissue increased with increasing incubation time. Perhaps, then, symbiotes export cholesterol to the host cell. Griffiths, G. and Beck, S. observed a vesicular system that may participate in this type of transport.

Oenocytes

Oenocytes are cells of ectodermal origin that are usually associated with the epidermis. In many species they also extend inward into the fat body where they are interspersed among the trophocytes. Oenocytes may be associated with the peripheral layer of fat body as well as with the perivisceral tissue. This distribution occurs in *Schistocerca gregaria,* with the peripheral location being more frequent. The association persists throughout all developmental stages including the mature adult. In old adults the oenocytes may contain yellow pigment and uric acid crystals. In adults of *Leptinotarsa,* oenocytes are found in fat body just below the integument at a density of up to two or three cells per fat body lobe. The structure of these oenocytes is similar to that of the permanent sub-epidermal oenocytes described by Locke, M.. They contain a large nucleus, few mitochondria and extensively developed tubular ER lacking ribosomes. As in trophocytes there may be small fields of glycogen. Trophocytes and oenocytes are intimately associated in *Leptinotarsa.* Although gap junctions have not been demonstrated, the plasma membranes of the trophocytes are as closely associated with those of the oenocytes as with each other. Oenocytes have also been described in the fat body of some Diptera. In *Sarcophaga bullata,* Stoppie, P. and co-workers observed that oenocytes and trophocytes were always separated by a basal lamina and no gap junctions were seen. Evans, J. has examined the fate of larval oenocytes associated with trophocytes at metamorphosis in *Dacus tryoni.* Both cell types undergo cytolysis when the larval fat body disintegrates at metamorphosis, and a new population of imaginal oenocytes differentiates in the adult fat body. The significance of this close and persistent relationship between oenocytes and trophocytes is not known. It is all the more interesting since it is not universal among insects. Romer, F. and co-workers have shown that isolated oenocytes of *Tenebrio molitor* synthesize ecdysteroids *in vitro*. Since fat body is known to convert ecdysone to 20-hydroxyecdysone and to respond to the latter, it is of interest to know more about the interrelations of steroid metabolism between oenocytes and fat body.

DEVELOPMENT

Polyploidy, Polyteny and Nucleolar Activity

In insects, tissue growth involving DNA synthesis and cell enlargement rather than cell division is common. It is found in tissues which must rapidly synthesize large amounts of a limited number of peptides. In fat body cells this growth pattern often precedes the synthesis of storage proteins during the last larval stadia of holometabola or vitellogenin synthesis in adult females.

Larval nuclear development

Inmost insect orders, except the Diptera, this type of nuclear growth is accomplished by endomitosis. In the fat body of both*Rhodnius prolixus* and *Calpodes,* the chromosomes replicate and the centromeres separate, but the nuclear membrane remains intact. For example in *Rhodnius,* Wigglesworth, V. reported that fat body cells are predominantly tetraploid at hatching, although a few octaploid cells are found. In fully nourished insects this condition persists throughout life although occasional 16n and 32n nuclei occur. Higher levels of ploidy, which result from cell and nuclear fusion, are found only after extreme starvation. In the early larval life of *Calpodes* the cells are tetraploid, and cell number increases by mitosis. At the beginning of the fifth stadium a period of DNA synthesis coupled with amitosis results in an increase both in ploidy and in cell number. Amitosis of the cells perpendicular to the plane of the fat body ribbon produces a tissue which is two cells thick. During the remainder of the stadium, the cells are predominantly octaploid although 16n and 32n cells are seen.

Polyteny is a special case of polyploidy in which the chromosomes undergo repeated cycles of replication without separation of the resulting strands. Polyteny, which is characteristic of the Diptera, is usually studied in tissues other than fat body, notably the salivary gland. An exception is the detailed report on the development of polyteny in the fat body of *Calliphora stygia*. *Calliphora* larvae hatching on day 0 ecdyse to the second instar 1 day later and to the third instar in the early half of day 3. At hatching, the fat body chromatin, which includes a highly condensed heterochromatin body, surrounds a prominent nucleolus. During the next 2 days the nuclei enlarge rapidly, and by the time the larva weighs - 2.5 mg, separate, irregular and elongate chromosomes can be resolved. At -48h after hatching polytene chromosomes can be recognized. The chromosomes contain short segments showing the characteristic banding pattern of polyteny. Between these are segments of diffuse, unbanded chromatin. The condensed heterochromatin body persists but is less conspicuous than at hatching. It apparently does not replicate as much as the rest of the genome. As growth continues with further cycles of replication, the chromosomes shorten and progressively occupy less of the nuclear volume. From the middle of the second stadium to the end of feeding on day 6 the rate of fat body protein synthesis is maximal. During this period further replication results in increasing chromosome width and banded segments continue to contract so that individual bands lie more closely together. Chromosomal replication ceases at the end of the feeding period. The banded sections of the chromosomes are thought to represent groups of reversibly inactivated loci while the intervening dispersed sections are analogous to the interphase chromatin of non-polytene nuclei. The segments which are dispersed during early larval life presumably include the loci involved in large-scale transcription prior to, or during, the peak period of protein synthesis.

Thomson has also described a relationship between developing polytene chromosomes and proliferating ribonucleoprotein (RNP) particles. During the first two stadia there is active RNA synthesis. The nucleoli enlarge and, by the time the period of maximum protein synthesis begins, there are multiple ribonucleoprotein particles of more or less uniform size. During the period of maximal protein synthesis these particles are associated with chromosomal segments other than the nucleolar organizers. Thomson suggests that the particles may participate in some aspects of processing, stabilizing or transport of gene products from the associated, transcriptionally active, chromosome segments. Further, this physical association may be a manifestation of a mechanism which couples mRNA synthesis with the synthesis of specific ribosomal populations. This relationship between RNP particles and specific segments of polytene chromosomes can be visualized because the latter are so large. There is, however, no reason to suppose that a similar relationship does not exist in polyploid nuclei. The

RNP particles near polytene chromosomes are presumably the same as the perichromatin granules that are believed to be packaged messenger RNA in other cells. From about 30 h into the last larval stadium of *Calpodes* the number of fat body perinucleolar ribosomal precursor particles and perichromatin granules increases, remains constant for about 30 h and then declines just prior to the intermoult period of massive protein synthesis. This increase in the number of RNP particles correlates with increased [^{3}H]uridine incorporation and the appearance of large numbers of ribosomes in the cytoplasm.

Adult nuclear development

The fat body cells of *Locusta migratoria* females are mostly tetraploid at ecdysis to the adult. These cells become octaploid and may even attain higher levels of ploidy at the start of vitellogenin synthesis. The ploidy values reported do not meet the predicted values for a doubling of DNA content during increasing ploidy, so that the interesting question of whether all the genome is replicated remains open. Allatectomy prevents the increase in ploidy but the application of a juvenile hormone (JH) analogue will induce it in allatectomized individuals. This JH-dependent polyploidy is thought to provide multiple copies of the genome to expedite the selective transcription of large amounts of mRNA specific for vitellogenin synthesis.

Although polyploidy is characteristic of the fat body cells of larval Diptera the nuclei of both young and old imaginal fat body cells of *Drosophila melanogaster* are the same size. They do not contain the thick, banded chromosomes seen in larvae.

Synthesis, Secretion and Storage— Structure in Relation to Function

Larval Syntheses

Preparation for synthesis

The fat body of the last larval instar has been the target of most studies on cell preparation for larval syntheses for two reasons. First, as the larva approaches its maximum weight the fat body weight increases proportionately and thus provides a convenient source of material for biochemical analysis. Secondly, most investigations have been primarily interested in describing the preparation for metamorphosis.

The synthetic activity of the fat body reflects the cyclical nature of growth and development. In *Calpodes* fat body, as in the epidermis,, there are intermoult/moult cycles of activity. In the fifth stadium both intermoult and moult are preceded by phases of preparation. The first is a period of preparation for larval syntheses that begins shortly after ecdysis at the fourth to fifth moult. This is followed by a phase of massive larval syntheses which ends with preparation for metamorphosis. A few hours after ecdysis the cells are small, with little cytoplasm, in which the mitochondria appear conspicuous. There is little RER and GCs are small with only a few transition vesicles and an incomplete stack of saccules lacking secretory vesicles (SV). During the first 2-1/2 days after ecdysis the cells become basophilic with many free ribosomes, an increasing amount of RER and larger GCs with the complete complement of three or four saccules. Lipid droplets arise and enlarge slightly as small areas of glycogen appear. This preparative phase is characterized by RNA synthesis. Autoradiography showed that incorporation of [^{3}H]uridine into cytoplasmic RNA increased until 66 h after ecdysis, just before the main period of protein synthesis. Incorporation of ^{32}P into ribosomal RNA of *Philosamia cynthia* was demonstrated by Takahashi, S. in early fifth-stage larvae in what was considered to be the main phase of fifthinstar rRNA synthesis. RNA synthesis during the preparative phase includes ribosomal and messenger RNA.

The accumulation of ribosomes and early stages of RER proliferation coincide with other syntheses for cell growth and metabolism. In *Calpodes,* preparation for fifth-stage larval syntheses includes the loss of fourth-stage peroxisomes and the formation of a new fifth-stage generation of peroxisomes from the RER. Peroxisomes are synthesized early in the stadium before the main secretory phase begins, and function through the rest of the intermoult.

The ultrastructural events associated with preparation for syntheses may be induced by nutritional changes. When fourth- or fifth-instar *Calpodes* larvae are starved for even a few hours, the cells undergo a resorptive phase during which lipid droplets are depleted, mitochondria fuse and mitochondria and RER are destroyed by autophagy. Refeeding is followed by a phase of mitochondrial division, completed within 4h, and proliferation of RER. In *Rhodnius,* mitochondrial destruction was observed after prolonged starvation of fourth instars, and refeeding resulted in synthesis of ribonucleoprotein, proliferation of RER, and mitochondrial recovery and division. Bosquet, G. showed that upon refeeding starved *Bombyx mori* larvae, the polysome level in the fat body increased 2.5-fold, thus dramatically increasing the protein synthetic capacity of the cells. There are many reports of protein synthesis preceded by RNA synthesis, often in relation to hormonal stimulation. Although few of these are cytological studies, the available data are consistent with the general principle that elaboration of the RER and GC are prerequisities for massive synthesis and secretion.

TABLE 6.2: CELL PREPARATION FOR CYCLES OF SYNTHESES

Order and species	*Observations Coleoptera*
Leptinotarsa decemlineata	Proliferation of RER, 1-2 days post-emergence. (A)t
Tenebrio molitor	RNA synthesis during the first 10 days shows cyclical pattern similar to protein synthesis. (A)
Dictyoptera	
Leucophaea maderae	Vitellogenin specific polysomes appear early in adult. ER-ribosome associations develop. (A)
Nauphoeta cinerea	RER proliferates 3-7 days post-emergence. (A)
Diptera	
Aedes aegypti	RER proliferation 8 h after blood meal. (A)
Calliphora erythrocephala	RER, mitochondria, free ribosomes increase in third instar. (L) RER proliferation, *1-2* days post-emergence. (A)
Calliphora stygia	Microsomal RNA synthesis increases after crustecdysone treatment in third instar. (L)
Glossina austeni	RNA synthesis at emergence precedes protein synthesis. (A)
Hemiptera	
Rhodnius prolixus	Starved fourth instars, refed, nucleolus enlarges, cytoplasmic RNA increases. (L)
Lepidoptera	
Antheraea pernyi	RNA synthesis increases early in adult development, and after injury to pupa. (P)

Bombyx mori	Starved fifth instars, refed, polysome level increased. (L)
Calpodes ethlius	RNA synthesis, RER proliferation mitochondrial division. (L) RER, mitochondria increase shortly after emergence. (A)
Hyalophora cecropia	RNA synthesis increases in early adult development and after injury to pupa. (P)
Philosamia cynthia ricini	Ribosomal RNA synthesis mainly in fifth instar. Uridine incorporation is highest on day 3 of fourth instar. (L)
Orthoptera	
Gryllus bimaculatus	RER proliferation 3-4 days post emergence. (A)
Locusta migratoria	Maximal uridine incorporation by day 8 of adult life. Polysomes specific for vitellogenin appear, RER and ribosomes increase. (A).

Phases of larval synthesis and secretion

The switch at E (ecdysis) + 66 h from the phase of cell preparation to larval synthesis in *Calpodes* is characterized by an increase in the amount of RER and the number and size of GC secretory vesicles together with increases in the rate of lipid and glycogen accumulation. The build-up of secretory protein in the intermoult is in the haemolymph, not the fat body. SVs are quickly secreted, suggesting that the control of exocytosis is not separable at this time from SV formation. These fat body SVs are the principal source of the large variety of haemolymph proteins secreted at all stages. In particular during the last larval intermoult they give rise to the haemolymph storage proteins. Storage proteins may also be synthesized in earlier stadia. They have been isolated and characterized for several species and are generally similar in size, composition and among the Diptera, immunological cross-reactivity has been observed. These storage proteins secreted into the haemolymph are later sequestered and stored in the fat body at metamorphosis. No larval function for these proteins has been identified other than accumulation of reserves. In *Chironomus,* the fat body of the last instar secretes haemoglobins which are regarded as primarily storage proteins although they may also function in oxygen

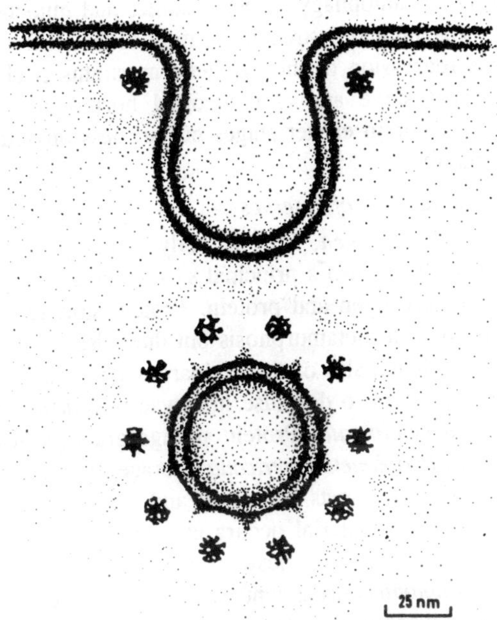

Fig. 6.1. The arrangement of the GC beads around transition vesicles. The beads are arranged in rings just above the smooth face of the RER where transition vesicles arise to form the Golgi complex. Transition vesicles pop through the rings which they leave behind them as they go on to become part of the outer saccule.

transport. Other haemolymph proteins secreted by the larval fat body include lipoproteins such as the diglyceride carrier, lipophorin, JH carrier protein and JH esterases. In *Trichoplusia ni* and in *Manduca sexta* there is a small peak of JH esterase activity immediately before the larval-pupal ecdysis. The fat body evidently retains its secretory ability at the time of metamorphosis, even when it is no longer primarily engaged in secretion.

Although there are no striking cytological correlates to this variety of syntheses for secretion, it is evident that not all products are secreted all the time, and the overall complement of secreted proteins depends upon the ambient controlling factors. Observations on *Calpodes* suggest that the changing pattern of some syntheses may be viewed ultrastructurally. Secretory vesicles with crystalline cores have been observed late in the fifth stadium but not at other ages. If these patterns are specific for vacuole contents, this may provide a marker for identifying proteins secreted at different times.

The end of the main secretory phase is characterized by a switch in GC structure with loss of SVs and an increase in 1° lysosomes. This is followed by autophagy of peroxisomes and mitochondria, and finally by the isolation and destruction of most of the RER in autophagic vacuoles. This regression of the synthetic apparatus is characteristic of the last larval stage and is controlled hormonally. Cycles of regression may occur in earlier stages but are less obvious than in a metamorphic moult.

Storage and mobilization of reserves

Protein, lipid, glycogen, The fat body is not only a source of secreted products but is also a principal storage organ. It may contain deposits of lipid, glycogen and protein. Protein storage granules are mainly associated with metamorphosis but they also occur in relation to gestation in adult females of the viviparous cockroach, *Diploptera punctata* and in relation to diapause in larvae of *Diatraea* and adults of *Leptinotarsa decemlineata*. Protein storage granules also occur in nymphal stages of *Periplaneta americana.* Storage of lipid and glycogen occurs at all ages, and reflects nutritional and metabolic requirements. Mobilization of stored material occurs in relation to moulting, adult syntheses, and starvation. Wigglesworth, V. observed that glycogen was depleted in *Rhodnius* during synthesis of chitinous cuticle, whereas proteins were mobilized during deposition of cuticular proteins. During starvation all reserves were mobilized, and a fully starved *Rhodnius* was depleted not only of reserves but also of mitochondria which are destroyed by autophagy. Starvation has similar effects on *Blaberus* and on *Calpodes.* Refeeding after starvation stimulates a burst of syntheses, organelles such as RER and mitochondria are quickly restored, followed by the formation of lipid droplets and other reserves. Although the biochemistry of lipid and glycogen synthesis and breakdown are well understood, little is known of the ultrastructural events during their deposition. The appearance and disappearance of these reserves have been frequently remarked upon, but little attention has been given to their genesis.

Tyrosine storage vacuoles

At certain times the fat body contains large fluid-filled vacuoles. They were first described in live *Chironomus* larvae by Voinov, V. who saw small vacuoles grow and fuse to form single large vacuoles in the centre of each cell. Wigglesworth, V. found vacuoles in newly ecdysed *Aedes* larvae and also in *Rhodnius* that had been fed after prolonged starvation. According to Wigglesworth, V. the watery vacuoles in *Rhodnius* fat body are derived from enlarged autophagic vacuoles that have lost their contents except for a rim of lipofuscin deposited around the inner membrane surface. The relation of these watery vacuoles to urate storage vacuoles or tyrosine storage vacuoles is therefore uncertain. In the fat body of *Leptinotarsa* larvae, vacuoles arise during the first stadium and decline at ecdysis. They then grow during the next three stadia before decreasing in size in the pupa. They are absent by the third day of adult life. Vacuoles also occur in the fat body of *Carausius morosus* and of *Calliphora* larvae. These large, intermittent structures have received surprisingly little attention. At ecdysis to the fourth instar in *Calpodes,* the vacuoles originating in the third instar occupy about 40% of the fat body volume. These vacuoles disappear within the next 6-12 h and a new population of fourth-instar vacuoles forms between 18 and 24 h after ecdysis. Brief (12 h) post-ecdysial starvation synchronizes their formation from the beginning of feeding (the technique can also be used to synchronize other kinds of cell behaviour such as mitochondrial division). The peak period for new vacuole formation occurs between 6 and 8 h after feeding. Surface infolds of plasma membrane form provacuoles which fuse to form vacuoles. By the further fusion and growth of these vacuoles each cell comes to contain a single giant vacuole 20-40 nm in diameter. They reach a maximum size at about 24 h before the end of the fourth stadium and then disappear abruptly in a 12 h period which centres on ecdysis to the fifth instar. In the fifth stadium the vacuoles are never as abundant as in the fourth, although they become a little more common prior to pupation. Vacuoles are not found during the pupal stadium or at the pupal-adult moult. The vacuoles are, therefore, a conspicuously larval, intermoult-moult-related phenomenon.

The vacuoles contain tyrosine and changes in the proportion of cell volume occupied by vacuoles correspond to changes in tyrosine content. The decline in fat body tyrosine levels and disappearance of the vacuoles at about the time of the fourth to fifth instar ecdysis correspond to an increase in haemolymph tyrosine concentration. Thus, fat body vacuoles release their stored tyrosine into the haemocoel during moulting to give a maximal concentration immediately after ecdysis. The disappearance of tyrosine from the haemolymph coincides with a need for phenolic precursors by the epidermis for cuticular tanning. Similar fat body vacuoles occur in *Leptinotarsa, Phormia, Manduca, Hyalophora* and *Chironomus,* suggesting that the storage of phenolic precursors, particularly tyrosine, in vacuoles may be a generality in insects.

Adult Syntheses

The fat body at emergence

The fat body of the adult is primarily a site of storage and synthesis of molecules which are mobilized for reproduction and flight.

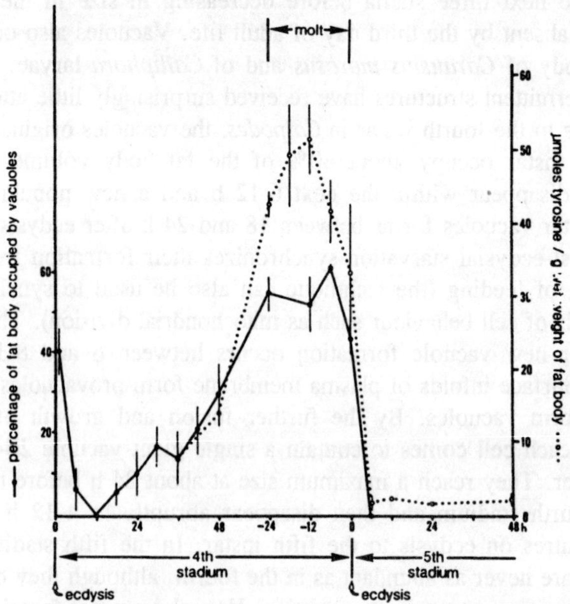

Fig. 6.2. Changes in the proportion of the fat body of Calpodes *occupied by vacuoles during the fourth and early 5th stadia (———) compared with the free tyrosine in the fat body (..........). The size of the vacuoles and the tyrosine content of the fat body go hand in hand.*

Ultrastructural studies of adult fat body have been predominantly concerned with development in the female, especially in relation to vitellogenin synthesis and secretion. At emergence, the cells have three notable features: large lipid droplets, extensive glycogen deposits and the residue of protein storage granules laid down in the premetamorphic stadium. The extent to which protein granules persist varies with species and sex. In most species in which the juvenile fat body is reorganized to form the adult tissue the storage granules persist throughout the pupal stage and are slowly depleted early in adult life. In silkmoths, development of the adult fat body coincides with ovarian maturation so that mobilization of the stored proteins and other preparative changes actually occur in the pupal stadium. In other Lepidoptera the granules persist for several days in the adult, while in cockroaches they may be retained throughout adult life though their numbers fluctuate with oviposition cycles. In the higher Diptera, the new tissue that develops after the larval fat body degenerates contains protein granules. Neither their origin nor fate have been described, although at some stage they contain hydrolytic enzymes.

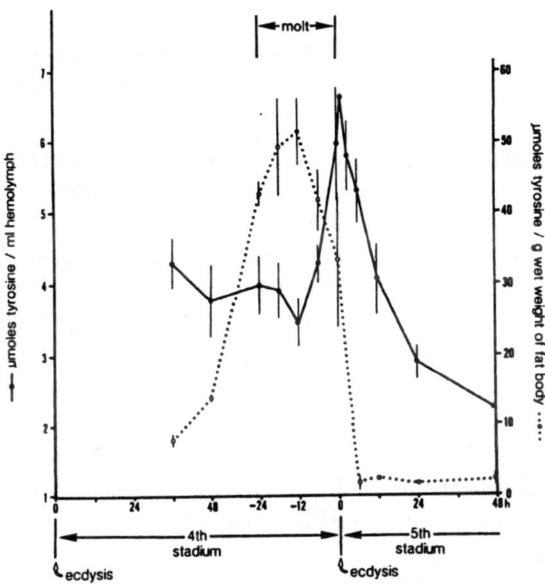

Fig. 6.3. The correlation between loss of tyrosine from the fat body and its elevation in the haemolymph during the fourth and fifth stadia of Calpodes. As the vacuoles decline in the fat body so tyrosine is briefly elevated in the haemolymph.

There is a marked sexual dimorphism in fat body ultrastructure that reflects the differing role of trophocytes in the reproductive process and sometimes in flight behaviour. Male adult fat body is characterized by relatively large lipid and glycogen stores and little ER. The cytoplasm is generally sparse and distributed between the large lipid droplets and in the perinuclear region. The large lipid store in male *Hyalophora* has been correlated with their energy requirements for very active flying in mate seeking. The overall impression is that the male tissue is not involved in massive synthesis of protein for export.

Preparation for syntheses

The ultrastructure of adult female tissue shows phases of activity which are remarkably similar to those in larval fat body. In the larva the phases are intermoult preparation, intermoult larval syntheses and

TABLE 6.3: ULTRASTRUCTURE OF FAT BODY OF PHARATE AND NEWLY EMERGED ADULTS

Order and species	*Description*
Coleoptera	
Leptinotarsa decemlineata	Cells contain glycogen, lipid droplets, protein granules, vacuoles. Little
Dictyoptera	
Nauphoeta cinerea	RER. (A)tMitochondria, lipid, glycogen, protein granules abundant. RER, Golgi
Diptera	
Aedes aegypti	complexes sparse. (A)Nuclear chromatin condensed, large nucleoli present. Mitochondria, lipid,
Calliphora erythrocephala	glycogen abundant. Some protein granules, free ribosomes. RER sparse.(A)Nucleus spheroid, large nucleolus. Mitochondria, glycogen, lipid, dense
Dacus tryoni	bodies, free ribosomes, present. RER sparse. (A)Mitochondria, lipid, glycogen evident. RER slightly distended. (PA)
Sarcophaga bullata	Lipid, glycogen abundant. Mitochondria, RER sparse. (A)
Lepidoptera	
Calpodes ethlius	Cells show steady loss of lipid, MVBs absent at E-40h, appear at
Galleria mellonella	E-30h, peroxisomes appear at E-24h, RER sparse. (PA)Autophagic vacuoles, protein storage granules with crystalline cores,MVBs peroxisomes, lipid droplets, dividing mitochondria. RER increasesby E+24h. (A)Autophagic vacuoles, lipid droplets present, protein storage granules
Hyalophora cecropia	partly degraded. (PA)Lipid, glycogen, present. Protein storage granules with crystalline cores
Orthoptera	
Gryllus bimaculatus	steadily degenerate. (PA)Mitochondria, lipid, glycogen abundant. Free ribosomes,

	protein granulespresent in females, absent in males. (A)Mitochondria, lipid droplets abundant, autophagic vacuoles present,
Locusta migratoria	RER, Golgi complexes sparse.Glycogen, lipid droplets, mitochondria often associated with lipid droplets, RER sparse. (A)Nuclei highly indented, large nucleolus, lipid, glycogen, protein granules (labelled lysosomes) present. (A)

cell remodelling for metamorphosis. In the adult these correspond to the previtellogenic phase, vitellogenin (Vg) synthesis and secretion, followed by the recovery or lysosomal phase.

The cells prepare for synthesis in the previtellogenic phase when the RER and GCs, poorly developed at emergence, quickly become abundant as the lipid and glycogen decrease. In *Locusta,* RER proliferation coincides with increases in total tissue RNA, uridine incorporation into RNA and the appearance of a class of polysomes specific for Vg synthesis.

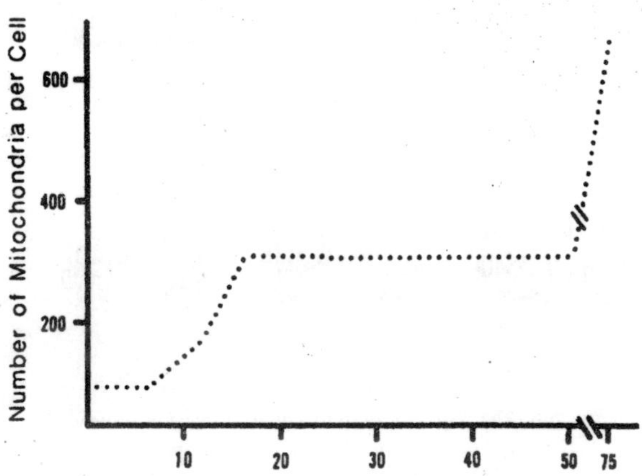

Fig. 6.4. The number of mitochondria per cell during intermoult preparation at the beginning of the fifth larval stage of Calpodes.

Vitellogenin synthesis and secretion

The vitellogenic phase is characterized by well-developed RER and GCs. The plasma membrane may be highly invaginated, and, in some species, e.g. *Leptinotarsa decemlineata* and *Calliphora erythrocephala*, there are relatively large secretory vesicles which, in the latter, have been shown to contain Vg. In *Locusta migratoria*, Vg was localized in RER cisternae, in GC vesicles and extracellularly in plasma membrane invaginations. The vitellogenic phase is closely linked in time with the development of terminal oocytes in the ovary and ends with oocyte maturation.

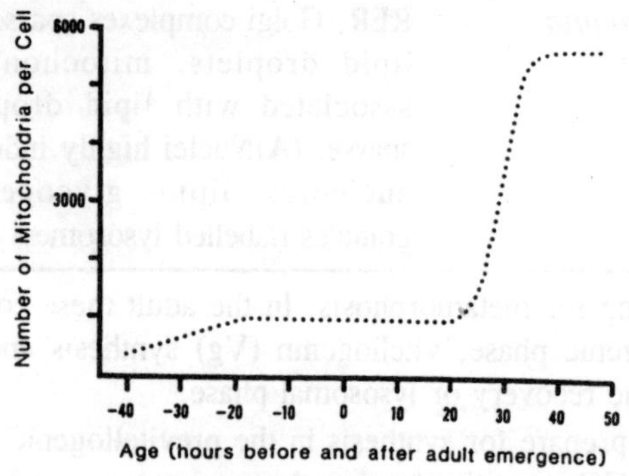

Fig. 6.5. The number of mitochondria per cell at the beginning of adult life in Calpodes.

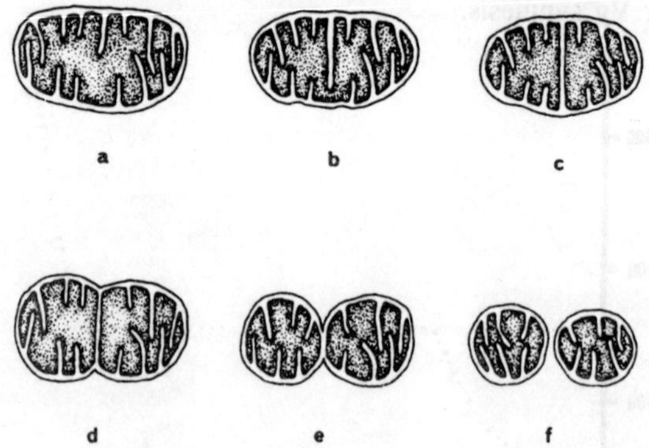

Fig. 6.6. The sequence of mitochondrial division in newly emerged adult Calpodes deduced from profiles

Post-synthetic recovery phase

The extent to which fat body cells are reorganized after termination of Vg synthesis varies. Some work clearly shows repeating cycles of activity. Rinterknecht, E. and Roussel, J.-P. reported a reduction in RER and an increase in free ribosomes. In *Locusta,* Couble, P. and co-workers described a distinct cycle associated with vitellogenesis which ends with the appearance of lysosomes. The content of the lysosomes is not clear, but bulk lysis of ribosomes seems unlikely since they persist for later Vg synthesis. Lauverjat, S., on the other hand, reported that after ovarian maturation the fat body is modified in a way that does not reflect the cyclic nature of vitellogenesis and subsequent egg-laying. Lysosomal structures were continuously present in older females. These observations may be reconciled by the variable extent to which the fat body is engaged in the cyclical synthesis of Vg and the continuous synthesis of other proteins. Haemolymph proteins such as lipophorin are also secreted during ovarian cycles of development but not in an overtly cyclic manner. Cellular change related to vitellogenesis is much more obviously cyclic in the Diptera where the fat body, as a consequence of the isolation and destruction of RER, reverts to a morphology similar to that seen in the previtellogenic stage.

Factors affecting cell structure related to vitellogenesis

Vitellogenin synthesis is controlled by more than one humoral factor. The influence of gland ablation and replacement or hormone therapy therefore varies between species. Some vitellogenin synthesis is clearly JHdependent. In *Locusta,* allatectomy inhibits Vg synthesis and the associated cytological changes, including those in the nucleus. Indeed, abnormal lipid accumulation rather than depletion follows corpora allata removal. Cell structure is rapidly returned to normal by implantation of active corpora allata or by the injection or topical application of JH or its analogues. In *Leucophaea,* JH stimulated the incorporation of labelled choline and orthophosphate into microsomal membranes *in vivo* and *in vitro*. This membrane synthesis was temporally coupled to synthesis of Vg. Although no ultrastructural evidence was presented, the results are consistent with the observations of Couble, P. and co-workers on the JH-dependent increase in RER in *Locusta*. Since vitellogenesis in *Locusta is* JH-dependent it is not surprising that ovary removal does not prevent the development of fat body structure appropriate for protein synthesis. In fact, Lauverjat, S. reported that this process was slightly accelerated when compared to controls. Further, the peripheral fat body cells, which Lauverjat claims do not normally undergo differentiation for large-scale protein synthesis, develop the same appearance as the periovarian cells after ovariectomy. Chen, T. and Wyatt, G. have shown that in *Locusta* Vg continues to be produced and accumulates to abnormally high levels in the haemolymph in ovariectomized females. Ovariectomy of *Leucophaea maderae* results in accumulation of Vg in the haemolymph and fat body and in *Leptinotarsa* also results in an increase in haemolymph Vg concentration. Surprisingly, in *Leptinotarsa* this surgery results in hypertrophy of the peripheral fat body cell type without any apparent increase in their protein synthetic machinery.

In the Diptera, Vg synthesis may be influenced by ecdysteroid secreted by the ovary because ovariectomy has a profound effect on vitellogenesis. In *Calliphora,* ovariectomy does not prevent the formation of a well-developed protein synthetic apparatus in the fat body, but there are no secretory vesicles in the GCs. Injection of 20-hydroxyecdysone counteracts the effect of ovariectomy. Immunocytochemical studies have shown that Vg is not present in the GCs and secretory vesicles of ovariectomized flies, but can be localized at these sites in controls and in adults injected with ecdysteroid after ovariectomy. In *Aedes aegypti* the role of ecdysteroids is more controversial. Behan, M. and Hagedorn, H. observed that in newly emerged adults the fat body cells show no signs of massive protein synthesis and secretion until several hours after a blood meal. The changes in ultrastructure, and the stimulation of Vg synthesis, correlate well with the known sequence of hormonal events that include secretion of the egg development neurosecretory hormone by the brain, and ecdysone secretion by the ovaries after the blood meal. However, Lea, A. has concluded that ovarian ecdysone does not stimulate vitellogenesis by the fat body even in blood-fed animals. On the other hand, Stoppie, P. and co-workers report that fat body in male *Sarcophaga bullata,* which does not normally generate large amounts of RER or plasma membrane infolding, can be induced to do so by the ingestion of moulting hormone that also induces subsequent Vg synthesis in males. In *Anopheles stephensi* also, 20-hydroxyecdysone stimulates Vg synthesis. In *Aedes* and *Calliphora* it is not clear what stimulates the development of cytoplasmic components necessary for large-scale Vg synthesis. Thomsen, E. and Thomsen, M. concluded that in *Calliphora* a factor from the medial neurosecretory cells is involved, although the evidence was not presented. The role of JH must also be questioned since the fat body must be exposed to JH before it becomes competent to respond to ovarian ecdysteroids. However, the cells show no evidence of major ultrastructural change following the natural period of JH exposure except for a slight alteration in the appearance of the RER.

One question which has so far remained unanswered is whether the oenocytes scattered among the trophocytes in the Diptera secrete ecdysteroids which might have a role in the control of ultrastructure. Since males of *Aedes,* and females after ovariectomy, contain about one-third the amount of ecdysone of normal-fed females, there must be another source of this hormone, possibly oenocytes. The conflicting results on *Aedes,* and on species other than Diptera, point to the need for the same careful assay of ecdysteroid titres and tissues of origin in adults as have been done on earlier stages.

There is some confusion in the literature as to whether the fat body in some species is regionally specialized for vitellogenesis. In *Locusta,* Lauverjat, S. claims that the peripheral fat body does not develop the necessary apparatus for Vg synthesis. In contrast, Couble, P. and co-workers claim no essential differences in fat body development in different body regions, and clearly state that peripheral fat body can secrete *Vg in vitro.* Similarly, de Loof, A. and Lagasse, A. suggest that peripheral fat body in *Leptinotarsa decemlineata* has an inappropriate structure for large-scale protein synthesis and yet, after ovariectomy which results in Vg accumulation in the haemolymph, only fat body cells having the structure of peripheral cells are found. It seems important that structural observation should not be over-interpreted. If fat body tissue can be shown to produce Vg *in vitro,* then structural arguments about how much RER or other cytoplasmic elements need be present for its synthesis become pointless.

Other Adult Syntheses

Since most studies of synthesis by adult fat body have been concerned with vitellogenesis, there is little information on other synthetic activities at this stage. There is evidence, however, that the adult fat body secretes a variety of haemolymph proteins. Indeed, even pupal fat body can be induced to secrete some of the same proteins that are made in larval and adult stages. Kramer, S. showed that increased JH esterase activity in adults of *Leptinotarsa* was stimulated by treatment with JH. Inhibition by actinomycin and puromycin suggested that the increased activity was due to synthesis. Esterase activity was also induced in pupae of *Hyalophora gloveri.* Lipophorin, a diglyceride carrier protein, has been isolated from the adult fat body of *Locusta* by Gellissen, G. and Wyatt, G., who showed that its production does not follow the same cyclical pattern as vitellogenesis. In adults of *Glossina austeni* labelled amino acids were incorporated into proteins secreted during larval development *in utero.* Protein granules were seen in the fat body during the first pregnancy, but neither their formation nor fate was described.

Structural changes related to adult diapause

In *Leptinotarsa,* females reared on short-day regimes enter diapause soon after adult ecdysis, whereas long-day females begin to oviposit within 5 days and continue to produce 50-60 eggs daily. During the first 2 days after adult ecdysis the fat body of pre-diapause and non-diapause females show no differences. At ecdysis both contain large vacuoles and numerous protein granules, some of which may contain urate. The cytoplasmic matrix is poorly developed. The protein granules decline rapidly in number and are absent by the second day. The vacuoles decrease and are missing by day 3 when differences in fat body structure between the two adult types become evident. In long-day females RER proliferates below the plasma membrane and adjacent to the nucleus, while increasing quantities of lipid appear between these two regions. The plasma membrane invaginates deeply and large numbers of mitochondria appear. In contrast, short-day females develop some RER and store much lipid after day 2, resulting in cell enlargement. On the 6th day of adult life protein granules begin to accumulate again and fat body cells swell to occupy most of the abdomen. By 10 days after adult ecdysis mixed autophagic/ heterophagic vacuoles appear and the beetles enter diapause 2 days later. These observations correlate well with data on protein synthesis. The fat body of both long- and short-day females synthesizes two vitellogenins and three diapauserelated proteins. Long-day females produce little diapause protein and large amounts of Vg, which are rapidly sequestered by the ovary. Short-day females produce small amounts of Vg and larger amounts of the three diapause proteins, two of which are returned to the fat body for storage. The autophagy, which starts before diapause inception, reduces much of the RER and many mitochondria to stored products for use in post-diapause development. The sequestered diapause proteins are thought to act as a nutrient store for diapause itself.

Fat body in allatectomized long-day females develops in the same manner as in untreated shortday females. This supports the contention that Vg synthesis is stimulated by JH, and that the production of protein storage granules is induced by its absence.

Cyclical Changes in the Cell Organelles

Golgi complex

The Golgi complexes (GCs) of insects are structurally less complex than those found in vertebrates or plants. In insects the GC at maximal development typically consists of a smooth face of RER bearings rings of GC beads and forming transition vesicles, an outer saccule, two or three inner saccules and according to the state of activity, secretory vesicles, 1° lysosomes and isolation envelope precursor vesicles. The

GC acts as a distribution centre for membranes from the RER and the materials they enclose. As a consequence, the appearance of GC profiles varies with the kind and amount of synthetic products passing through them. The route that these materials take through the GC depends on their fate. There are at least three major processes. These are the processing and packaging ofproteins in secretory vesicles for export to the haemolymph, the production of 1° lysosomes which contribute lytic enzymes to both autophagic and heterophagic vacuoles, and the formation of isolation membranes for autophagy. Isolation membranes are probably formed from vesicles derived from the outer saccule of the GC because both show pronounced osmiophilia and stain with lead. The isolation membranes themselves do not carry the lytic enzymes required for organelle or protein breakdown. These enzymes are transported in primary lysosomes arising from the inner saccules of the GC, as suggested by the demonstration of acid phosphatase activity in both structures. In *Calpodes,* large secretory vesicles containing protein for export form from the innerGC saccules. Large-scale protein secretion does not always correlate with the presence of large secretory vesicles. For example, vitellogenesis in *Calliphora vicina* coincides with the presence of large compound secretory vesicles in the cytoplasm, whereas in *Locusta* no Vg transport between the GC and the extracellular space could be found. The difference may be due to the rate of movement of SVs from the GC to the surface and the coupling to exocytosis. Control of the rate of exocytosis may be separate from the control of the rate of protein synthesis and, in *Locusta migratoria,* protein may be exocytosed rapidly.

A unique staining property of arthropod GCs has permitted the discovery in fat body cells of a subcellular structure found in all GCs. The smooth surface of the RER near GC elements has bead-like particles arranged in rings at the base of forming transition vesicles. In arthropods, but not in other phyla, these GC beads stain specifically with bismuth salts. They have a diameter of 10-12 nm and are separated from each other and the membrane by a clear halo, giving them a centre-to-centre spacing of 27 nm and a centre-to-membrane spacing of 14 nm.

Transport of secretory proteins and membrane between RER and GC saccules is accomplished by the transition vesicles in an alternating fission-fusion process. This transport is sensitive to inhibitors of oxidative phosphorylation which block the energy-dependent step at the forming face of the GC, between the smooth surface of the RER and the transition vesicles. Brodie, D. studied the effect of a variety of inhibitors of protein synthesis and oxidative phosphorylation, and the effect of ionophores on bead ring integrity. He found that bead ring integrity is independent of protein synthesis and of late GC processing of secretory protein. Of the nine experimental treatments used, only those that lowered intracellular ATP levels altered ring morphology. When ATP levels were reduced the beads clumped together and lost their ring structure, although no change in centre-to-centre spacing or centre-to-membrane distance occurred. The concentration of beads into a close-packed sheet may be caused by a loss of membrane, which reduces the area of the forming face of the GC. This study suggests that bead ring integrity is ATP-dependent. Further, the block to transition vesicle formation concomitant with the collapse of bead rings suggests that formation of transition vesicles is dependent upon the integrity of the rings. The beads may therefore be the structural correlate of the energy-dependent gating mechanism controlling transport in the ER-GC transition region described by Jamieson, J. and Palade, G..

The size of the GC varies with the secretory activity of the cell. At times of low secretory activity the GCs may be barely recognizable, which perhaps accounts for their seeming absence in some older ultrastructural studies. In *Calpodes* no stage can be identified when the GC beads are absent. The minimal structure is one or a few rings of beads adjacent to a smooth surface of ER, as found during GC formation at the beginning of intermoult preparation early in the last larval stadium in *Calpodes ethlius*. The GC beads are prime movers in the differentiation of GCs from the RER.

Peroxisomes

In *Calpodes,* mature peroxisomes are flattened with a depression on one side. A dense core of coiled tubules is connected to the limiting membrane at the depression by a finely granular mass. Peroxisomes contain catalase and urate oxidase. In the fourth larval stadium the peroxisomes decrease in size before ecdysis to the fifth stage. During the day following ecdysis all the old perioxisomes atrophy while a new generation appears. Peroxisomes form as diverticula of the RER, and grow by the addition of material from confronting RER cisternae. The mature peroxisomes become progressively smaller during the phase of intense fat body syntheses and are isolated and lysed before pupal ecdysis. Pupal fat body contains no perioxisomes until about 24 h prior to adult emergence, when they arise and grow in the same manner as in larvae. Although adult peroxisomes contain a coiled core it is never as well developed as that in larvae. The presence and degree of complexity of the core depends on fixation temperature. Coiled structures present after fixation at 0-4° are not seen after fixation at room temperature, suggesting that they are protein liquid crystals. In both larvae and adults there is often an association between peroxisomes and lipid droplets. Peroxisomes occur in the fat body of other insects such as *Periplaneta* as well as in other tissues, but they have been little studied.

Mitochondria

As a consequence of mitosis at the end of the fourth stadium in *Calpodes* there are fewer mitochondria per cell after ecdysis to the fifth stage. However, during a 10 h period between E + 6 and E + 16 h the number of mitochondria per cell increases from about 100 to 300. Detailed stereological analysis showed that the increase in number results from mitochondrial division which follows a period of rapid growth and shape change. A further round of replication at E + 17 h increases the number of mitochondria to ~700 per cell . Thus, during the period of greatest fat body synthetic activity (E + 66 h to E + 156 h) a large population of mitochondria is present to supply metabolic needs. After this period, most larval mitochondria are isolated and destroyed so that few remain during the pupal stadium. At emergence mitochondrial number rises again to culminate in a dramatic 6-fold increase between 22 and 34 h after adult eclosion. Mitochondrial division has a characteristic morphological sequence both in the larva and adult. It begins as one crista grows across the mitochondrial matrix' and fuses with the opposite inner membrane, thereby separating the matrix into separate parts. A constriction forms at the site of the transecting crista which tightens to separate the mitochondrion in two. The mitochondria of larval fat body, including those which persist into the early pupal stage, are characteristically rod-shaped, while those of late pupae and adults tend to be spherical or oval. In larval fat body the rodshaped mitochondria first assume a spherical form before dividing. A brief study of *Calliphora erythrocephala* showed differences in mitochondrial morphology in feeding and post-feeding larvae. During feeding, mitochondria) profiles were irregular with a faint matrix; they later became smaller with a more regular outline and dense matrix. A morphometric study of pre-metamorphic changes in the fat body of *Calliphora* showed a similar pattern. At the start of the last larval stage, fat body mitochondria were irregular and elongate but as the end of feeding approached they became more rounded or oval. The absolute area of mitochondria increased throughout this period but dropped by 50% when feeding ceased, presumably as a result of a round of autophagy occurring at this time.

The change in form during development suggests a kind of mitochondrial differentiation. There may also be morphological differentiation of the cristae, which can sometimes be tubular rather than lamellate. Do these differences reflect changed functions and enzyme compositions? For fat body mitochondria these questions remain to be answered. An investigation on the fat body of *Aldrichina grahami* attempted to correlate developmental changes in mitochondrial size and morphology with the activity of three enzymes from mitochondrial fractions. During feeding, the mitochondria showed a greater average diameter in thin sections, were more electron-dense, and were claimed to be more numerous than in older larvae. In postfeeding larvae both large and small mitochondria were found. Early in the stadium, the fraction of large mitochondria accounted for almost all of the activity of three mitochondrial enzymes. After feeding ceased most of this activity was associated with the cell fraction containing small mitochondria.

Little is known about factors controlling mitochondrial morphology. In *Rhodnius,* starvation results in mitochondrial fusion so that each cell comes to contain only a few giant mitochondria. After feeding, the giant mitochondria divide to produce many normal organelles. In *Calpodes* giant mitochondria also result from even a brief starvation. These divide to become normal again within 4 h of feeding. Clearly, then, nutrition can influence mitochondrial size and division. It is less certain that it influences mitochondrial growth, DNA replication and division during normal development. Both starvation and a ligature placed posterior to the prothoracic glands prevent any mitochondrial genesis in the first 26 h of the last larval stadium. It is difficult to separate the effects of starvation from possible influences of anterior endocrine centres. However, the coincidence between the timing of the two phases of mitochondrial replication and two significant elevations in haemolymph ecdysteroid titre early in the last larval stadium of *Calpodes ethlius* suggests a proximal endocrine influence worth further study.

Regional Differences in Fat Body Structure and Function

There is considerable evidence that trophocytes, at least in some insects, show regional diversity of structure and function. Differences have been observed at all stages of development and in several taxonomic groups. Several authors have described ultrastructural and functional differences between the peripheral and visceral regions of larval and adult tissue, and have claimed that there are differential regional responses to a variety of conditions including different hormonal milieux. The evidence suggests that in several species the peripheral tissue is primarily a site of lipid synthesis and storage, while the visceral tissue is specialized for protein synthesis and storage, and for heterophagy associated with storage. However, this dimorphism is not consistent, as the case of *Chironomus* shows.

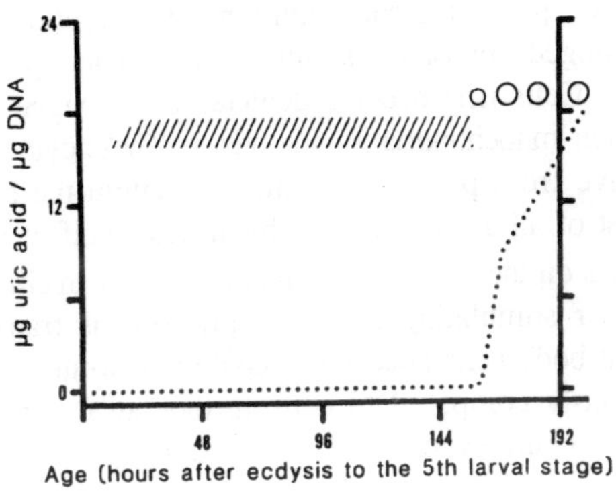

Fig. 6.7. The relation between the disappearance of peroxisomes (/////), the accumulation of urate in the fat body (......) and the accumulation of urate storage granules (O O) in Calpodes.

The clearest case for regional specialization has been made in the Diptera where there are two kinds of differentiation. The first is a functional differentiation of the peripheral and visceral tissues. Schin, K. and co-workers studied the timing and localization of haemoglobin (Hb) synthesis in fourth-instar larvae of *Chironomus thummi*. Incorporation of [^{14}C]aminolevulinic acid, and localization of peroxidatic activity of Hb by the diamino benzidine reaction, showed that Hb synthesis occurred in the peripheral (sub-epidermal) tissue. Visceral tissue was principally involved in pinocytosis of haemolymph proteins, including injected ferritin, and at times, the autophagy of cell organelles.

A second kind of tissue variation observed in the Diptera is an anterior-posterior gradient of development which occurs at metamorphosis. In the last larval stage, fat body cells sequester and store blood proteins in granules. In *Drosophila* the formation of storage granules begins in the anterior region of the fat body and proceeds posteriorly. A similar gradient has been described for the cell separation that precedes histolysis in *Sarcophaga*, *Phormia* and *Drosophila*. The pattern of lysosomal activity and autolysis is similar in *Calliphora*.

Regional differences in the rate of tissue autolysis have been described in other groups, though not with an anterior-posterior gradient. In *Leptinotarsa,* the peripheral and some visceral cells are lysed, whereas the dorsolateral group of visceral cells persist into the adult stage and are thought to give rise to the adult fat body. Waku, Y. and Sumimoto, K. concluded that there is differential autolysis of the visceral cells of *Bombyx mori*.

Gradients of autolysis and other cytological changes occur in response to such physiological conditions as starvation. In *Rhodnius* nymphs, prolonged starvation results in increasing polyploidy after nuclear fusion. Cells first become depleted of glycogen and lipid and then destroy their mitochondria by autophagy. These cytological changes occur in a wave from posterior to anterior within the abdomen, and are evident last of all at the sides of the anterior half of the abdomen.

Experiments on the control of increase in ploidy in *Calpodes* suggest that some factor stimulating nuclear replication is transmitted from cell to cell. Fat body cells attached to epidermal wounds may increase ploidy enormously compared to cells further away. Presumably the gap junctions are involved in all such examples of linearly graded differentiation.

Functional differences between fat body cells from different parts of the animal have been reported in relation to metamorphosis and adult syntheses. In a study of RNA synthesis during diapause and adult development in silkmoths, Berry, S. and co-workers, observed that the perigonadal fat body incorporated uridine at a much higher rate than the rest of the fat body. In *Locusta* Lauverjat, S. characterized the peripheral fat body of adult females as having a male phenotype. These cells are not ultrastructurally differentiated for protein synthesis and secretion, and show no cylical activity. Visceral fat body, especially near the ovaries and gut, has welldeveloped RER and GC, the abundance of which varies cyclically with secretion of Vg during ovarian maturation cycles. As observed by Lauverjat, S., the cells of these two regions showed different responses to ovariectomy and allatectomy. Visceral cells lost their protein secretory ultrastructure and assumed the appearance of the peripheral cells that store lipids. These observations on regional differentiation have not been confirmed by others working with *Locusta,* possibly because of differences in the timing of their observations or other variables in the studies. The two cell types described by Lauverjat for normal females were observed by Rinterknecht, E. and Roussel, J.-P. in the visceral fat body of

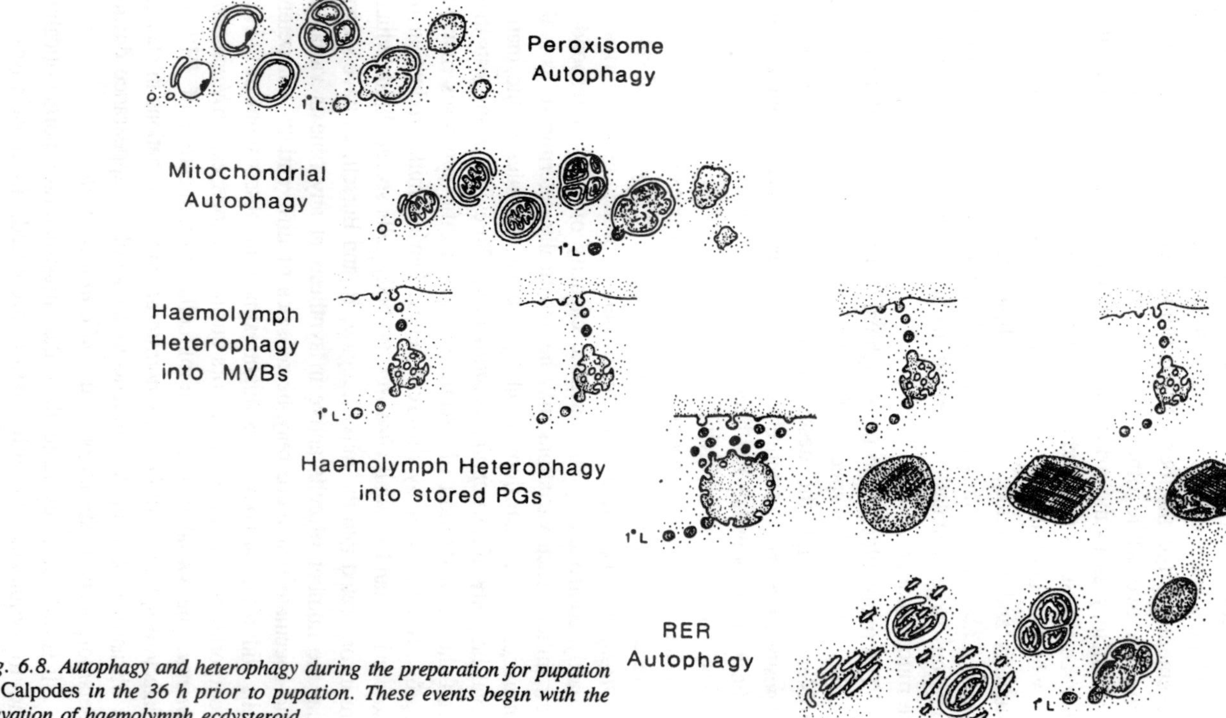

Fig. 6.8. Autophagy and heterophagy during the preparation for pupation in Calpodes *in the 36 h prior to pupation. These events begin with the elevation of haemolymph ecdysteroid.*

allatectomized females but Couble, P. and co-workers saw no ultrastructural differences between peripheral and visceral tissue in normal or allatectomized females. In *Leptinotarsa* also there is no agreement on the number of functional cell types, some authors recognize two while others propose three. The evidence is also contradictory as to the principal site of Vg synthesis and secretion; the visceral tissue is the candidate favoured by de Loof, A. and Lagasse, A., whereas Dortland, J. and Esch, T. found evidence for protein synthesis and secretion, as well as lipid storage, in the peripheral tissue. Labour, G. showed that peripheral fat body cells contain numerous large vacuoles, whereas in the perivisceral tissue the vacuoles are much smaller.

The many allusions to topographical and functional differences in the fat body of the species described above reveal an unresolved problem. Although the descriptions provided by each author are suggestive, they are not entirely convincing, largely because clarification of these distinctions has not been a principal objective of any of these studies, except those involving Diptera. The evidence from this order provides convincing evidence for unequal competence of cells in different regions to respond to particular developmental and physiological stimuli. In *Drosophila,* the antero-posterior gradient of protein granule formation is partly controlled by the intrinsic capacity of cells to respond to their milieu. Such variations may be genetically determined, as is the formation of the autofluorescent kynurenine granules in the mature larval fat body of *Drosophila melanogaster.* The evidence suggests that even the incorporation of precursors for specific syntheses may be regionally controlled by different genes. During milk production by the uterine gland in adult *Glossina,* fat body cells secrete lipids which are incorporated into the milk. Langley, P. and Bursell, E. have shown that the number of cells active in synthesis at any time is variable, and speculate that there may be mosaics of differentially competent cells within the fat body. This might explain the observations of uneven incorporation of labelled precursors during vitellogenesis in *Nauphoeta cinerea*, the variable activation of cells after injection of JH into allatectomized adults of *Locusta migratoria,* and the variations in change of ultrastructure from a synthetic to an autolytic appearance during metamorphosis of the fat body in *Galleria mellonella*.

It is difficult to evaluate all of the ultrastructural reports critically, and to synthesize the information presented. In most cases the descriptions of regional or temporal variation have not been clearly illustrated by micrographs. In addition, with few exceptions, no attempt was made to standardize the regions from which tissue was taken for study. Thus temporal variations could have been mistaken for regional ones. Systematic analysis is required before a more thorough synthesis can be attempted, and the question of the existence and significance of some of these putative differences must remain open.

METAMORPHOSIS

Metamorphosis in the fat body involves a series of events that result in significant changes in cell and tissue structure. Several characteristic processes have been described in the Holometabola, though not all of these occur in every species. The main events are: heterophagy and storage of haemolymph proteins; autophagy of some cell organelles; cell separation followed by tissue and cell remodelling; and histolysis of the larval tissue and replacement by a new adult tissue.

Heterophagy and Storage of Haemolymph Proteins

Larval fat body of many species takes up proteins from the haemolymph by pinocytosis. The proteins enter in small vesicles which fuse with each other to form larger granules. In *Calpodes*, pinocytosis occurs throughout the fifth stadium. During the feeding stage the proteins are hydrolysed in multivesicular bodies (MVBs), which are a constant feature of all or most cell types because of their role in membrane turnover. During metamorphosis there is a switch from lysis to storage, and the sequestered protein is stored in large membranebound protein granules. Pinocytosis of haemolymph proteins into granules occurs during the pre-metamorphic phase of last instars of *Calliphora erythrocephala* and *Drosophila.* Although these granules were not identified ultrastructurally, they are probably MVBs since in other structural and developmental characteristics these cells resemble those of *Calpodes*. Tobe, S. and Loughton, B. failed to find MVBs in fat body of fifthinstar *Locusta,* even though there was significant uptake of labelled blood proteins. Coupland, R. did not see protein granules in fat body of first to fifth instars of *Schistocerca gregaria.*

However, Lauverjat, S. showed crystalline protein granules in the fat body of newly emerged females of *Locusta.* The absence of protein granules in fifth instars may reflect a very rapid turnover of pinocytosed proteins at this stage, and a much later switch from lysis to storage. In *Periplaneta* both recently emerged fifth-stage nymphs and adults have crystalline granules of stored protein.

TABLE 6.4: SELECTIVE UPTAKE AND STORAGE OF HAEMOLYMPH PROTEINS AT METAMORPHOSIS

Order and species	*Method*	*Observations*
Coleoptera		
Leptinotarsa decemlineata	EM	Storage granules, fused autophagic vacuoles and storage granules.
Diptera		
Calliphora erythrocephala	EM	Storage granules, fused autophagic vacuoles and storage granules.
Calliphora stygia	EP	Uptake and storage of calliphorin (protein C) from haemolymph.
Drosophila melanogaster	LM	Storage granules in last instar larvae.
	LM,Ce	Storage granules isolated.
	EM	Storage granules, autophagic vacuoles.
	IP,EP,EC,EM	Storage protein, drosophilin, isolated.
Lepidoptera		
Bombyx mori	Ce,Ch,EP, ID,IE	Storage granules, autophagic vacuoles, fused storage
	IE,EM	granule and autophagic vacuoles.
	EM,LM	Storage (albuminoid) granules, urate granules, autophagic vacuoles.
Calpodes ethlius	Ch,EP,FI,ID	Uptake of 3 antigens 470 K, 580 K, 720 K, M.W.
	EP,FI	Uptake of two proteins, 580 K, 720 K.
	EM,LM	Storage granules, autophagic vacuoles, fused storage granules and autophagic vacuoles.

Diatraea grandiosella	EP	Storage of three haemolymph proteins.
Ephestia kuhniella	Ce,EC,EP,IE,	Storage of two haemolymph proteins.
	EM,LM	Storage granules.
Galleria mellonella	EP,ID	Uptake of three haemolymph antigens.
	EM	Storage granules, autophagic vacuoles.
Hyalophora cecropia	Ce,EP,ID,EM	Storage granules, containing two haemolymph proteins,
	LM	urate granules.
	EM	Storage granules.
Malacosoma americanum	ES,ID	Several haemolymph antigens in late larval and pupal fat body.
Manduca sexta	Ch,Cm,EP	Haemolymph protein, manducin occurs in pupal
	ID,IE	fat body.
Philosamia cynthia ricini	EM	Storage granules.
	EM	Storage granules, autophagic vacuoles.
Pieris brassicae	EP	Two haemolymph proteins stored.
Protoparce quinquemaculata	ES,ID	Uptake of two haemolymph antigens.
Orthoptera		
Locust a migratoria	EM	Storage granules (lysosomes) at imaginal ecdysis.

The appearance of storage granules varies with larval and pupal age. They become increasingly crystalline as the pupal stage progresses, and in some species they are almost completely crystalline at adult emergence. They may also appear heterogeneous in their content as the granules of pinocytosed proteins fuse with autophagic vacuoles containing mitochondria and RER. Such mixed granules may also develop crystalline regions. Crystallization may have particular significance in relation to the number and kind of haemolymph storage proteins that have been pinocytosed. Does crystallization imply that only one protein species survives in a storage granule, for example?

Selective Uptake of Haemolymph Proteins

The formation of storage granules from sequestered haemolymph proteins was described in *Calpodes* by Locke, M. and Collins, J.. Although the original demonstration of uptake involved the use of foreign proteins injected into the haemolymph, the process is normally selective and only specific proteins are sequestered. *Tojo,* S. and coworkers isolated the storage granules from last-instar larvae and pupae of *Hyalophora cecropia,* and identified their contents as the two major protein fractions of the haemolymph of the last instar. Storage proteins have also been isolated from pupal fat body of *Bombyx mori*, and are presumed to be stored in the protein granules of that tissue. The existence of storage proteins and their sequestration in granules in fat body are widespread phenomena, but little is known of the cellular basis of selective uptake. Most storage proteins are thought to be hexamers with molecular weights near 500K daltons though they may be as large as 900 K daltons personal communication). It is not known whether they are taken up as monomers, nor is it obvious how particles the size of the polymer would cross the basal lamina to enter the cells by selective endocytosis. Studies on the molecular mechanisms of selective uptake have been impeded by the difficulty of isolating pure storage proteins from haemolymph and fat body. These problems have been overcome for several species of Lepidoptera, e.g. *Manduca, Bombyx,* and *Hyalophora*. Kramer, S. and co-workers have evidence that the chromatographic, electrophoretic and immunological properties, as well as the amino acid composition, of manducin are the same in the larval haemolymph and the pupal fat body of *Manduca sexta*. This suggests that there are no significant molecular changes during uptake that might mediate the selective process, as does the removal of sialic acid in vertebrate liver.

Sex Differences in Protein Storage

The amount of protein stored in granules in the fat body may vary with sex. In *Calpodes* pupae the proportion of cell area occupied by granules was greater in females than in males. Striking differences between sexes were reported in *Bombyx* by Waku, Y. and Sumimoto, K.. These were quantified by Tojo, S. and co-workers who reported that in female pupae 60% of the total fat body protein was in storage granules compared with only 20% in the male. Similar sex differences were seen in *Hyalophora cecropia* pupae, and Bhakthan, N. and Gilbert, L. found no protein granules in adult males. Male silkmoth pupae also differ from females in containing significantly more urate/protein granules.

Urate Storage

Uric acid is stored in the fat body of many species in membrane-bound compartments varying in size from the giant vacuoles of urocytes to the small MVB-like granules of Lepidoptera. In *Hyalophora cecropia* it begins to accumulate after cessation of spinning, and correlates with an increasing accumulation of granules showing a characteristic fibrous internal structure. These granules, which are much more numerous in males, have been isolated and shown to contain 75% uric acid and about 24% protein. Unfortunately the granule fraction was too hard to section but its apparent homogeneity under the microscope led these authors to suggest that the protein is part of the granule and not a contaminant. Granules with a similar fibrous matrix are evident in micrographs of fat body from other silkmoths. In *Calpodes,* granules with a fibrous matrix develop at a time when uric acid accumulates in the fat body. The development of these granules overlaps the switch from the formation of lysosomal MVBs to the formation of protein storage granules at the start of metamorphosis. This also coincides with the loss of urate oxidase activity by the fat body when the peroxisomes are destroyed. The stored urate in the granules presumably arises from the metabolism of nucleic acids that accompanies the end of massive protein syntheses and the autophagy of the RER. There may also be uptake of uric acid from the haemolymph as observed in *Manduca* by Buckner, J. and Caldwell, J.. The origin of the protein component of the urate granules may also be extracellular as urates may be transported as proteinbound complexes. Tojo, S. and co-workers concluded that the urate granules contain proteins distinct from the proteins in the storage granules. In *Calpodes,* vacuoles having a fibrous matrix show a developmental sequence that suggests an origin from a MVB-like structure. Initially there are MVBs with dense contents of a slightly fibrous appearance. Later stages show less dense, more fibrous contents, and still later, the vacuoles appear almost empty except for the accumulated fibres. These fibrous vacuoles were identified by Locke, M. and Collins, J. as residual bodies since they may also contain acid phosphatase. The development sequence described above suggests that urate granules may arise partly by endocytosis, with endogenous uric acid being added to the vacuole of endocytosed protein.

Larsen, S. showed that in late pupal *Calpodes* MVBs may develop into granules with a fibrous matrix. When thorotrast, a non-degradable tracer, was injected into the haemolymph it was first localized in pinocytic vesicles, and later in what appeared to be residual bodies (probably urate granules). Urate granules in the late pupa *of Calpodes* may therefore be vacuoles containing haemolymph proteins pinocytosed at different times, together with uric acid and lysosomal enzymes. Isolation of the granules, and analysis of their contents, should clarify their relationship to residual bodies, and their role in the changes in nitrogen metabolism from larva to pupa.

Autophagy of Cell Organelles

Early in metamorphosis the fat body switches from activities concerned with synthesis and secretion to the storage and reutilization of reserves during the pupal and adult stages. This change involves removal of some organelles and their subsequent replacement. Studies on *Calpodes* have shown that components to be digested are first isolated from the cytoplasm and fuse with 1° lysosomes to digest the contents. In this way all peroxisomes, and most of the mitochondria and endoplasmic reticulum, are destroyed in autophagic vacuoles within which these organelles are clearly identifiable soon after their isolation. As lysis proceeds, organelles lose their integrity and gradually degenerate into dense amorphous granules. This organelle-specific sequence of destruction shows that cells have a mechanism for selecting particular cell components for autophagy. In fat body at least, organelle turnover is not random. Autophagic vacuoles may also fuse with each other and with protein storage granules to form structures of heterogeneous appearance and origin. This has been reported for many species, as shown in Table 4. Autophagy results in complete or partial destruction of organelles, preparatory to a change in cellular activity. Although some RER persists, the cells are principally engaged in storing protein, lipid, glycogen, and in some species, urate.

Formation of the Adult Fat Body

Cell and Tissue Reorganization

Two mechanisms have been described for the origin of the adult fat body. It may develop by reorganization of the larval tissue, or differentiate from stem cells carried in the pupa. Both mechanisms have been described among the Diptera. In *Aedes aegypti,* larval tissue is thought to persist in the adult, whereas Evans, A. and many subsequent authors have claimed that in the cyclorrapha there is complete destruction of the larval tissue. The process begins with separation of the larval cells from each other, and the breakdown of the basal lamina. The cells appear to die early in the pupal stage and by the third day of adult life the larval tissue is no longer recognizable. Butterworth, F. chronicled the disappearance of larval tissue in *Drosphila,* but the process was not illustrated. In *Dacus tryoni*, Evans, J. has described the origin of imaginal fat body from groups of cells that appear among imaginal oenocytes. They were said to arise from the new adult epidermis that develops from groups of stem cells carried in the pupa. These accounts of the fate of larval fat body and the origin of the adult tissue in the cyclorrapha deserve further examination. Rizki, T. has described the arrangement of larval cells in the larva and early adult of *Drosophila melanogaster*. At hatching the number of larval fat body cells is fixed and there is no further cell division. Following pupariation the fat body is reorganized so that the cells occupy new positions in the pupa and adult. The abdominal cavity, thorax and head capsule of the adult contain larval cells which float out freely into the bathing medium during dissection. These cells are frequently associated with haemocytes, loosely attached to their surfaces. Observations on a redpigmented mutant have cast some doubt on the claim that the larval fat body is completely replaced by a new adult tissue. Rizki described the localization of red-pigmented larval cells in the head capsule of newly emerged flies of the *tu-W-rc* mutant. Later, pigmented elements were observed in the adult tissue, but he was unable to distinguish whether these were transformed larval cells, or adult cells that had incorporated larval elements. This raises the question of whether, and to what extent, larval cells contribute to the formation of adult tissue. The view that larval fat body cells die in the adult has not been supported by microscopical evidence. Several authors (for example Butterworth, F.; Thomsen, E. and Thomsen, M.; and Evans, J. refer to the disappearance of the larval cells in the adult, but there are no illustrations of the process.

The origin of the adult tissue is equally uncertain. Evans, J. concluded that in *Dacus* it develops from the new epidermis. This view conflicts with the embryonic origin of the larval fat body from mesoderm. Although Anderson, D. identified the imaginal discs from which the adult epidermis develops he gave no indication of mesodermal rudiments from which fat body might develop. It is difficult to see how the adult tissue could arise from the epidermal discs, unless the concepts of classical embryology have no relevance to this group of insects. The reorganization of the larval tissue in the pupa and adult, the association with haemocytes, the lack of microscopical evidence of cell death, and the absence of an identified stem cell population, all indicate that not enough is known about the metamorphosis of the fat body in the cyclorrapha. Although there are clear descriptions of the beginning of the metamorphic process there is little ultrastructural evidence for later stages. We know of no evidence on the fate of larval fat body cells, or the origin of the imaginal cells, which could not be interpreted as an extreme example of cell and tissue remodelling.

In other orders the remodelling of larval cells into adult tissue has been described in greater detail. However, because of the difficulties experienced by some investigators in obtaining adequate fixation of the tissue, there are some confusing conclusions in the literature. Larsen, S. described the process in *Calpodes,* identifying three characteristic stages. As in the Diptera, the cells lose their intercellular connections shortly before pupation, but they remain as a tissue enclosed by a basal lamina. Later, they become clumped around tracheoles, losing their sheet-like arrangement by 2 days before adult ecdysis. The reorganized cells re-establish intercellular connections, as shown by the lanthanum permeation of gap junctions. These changes in cell and tissue reorganization do not involve cell death or cell division. Haemocytes may be involved in the rearrangment of cells. Walters, D. observed that during metamorphosis in silkmoths, cells which had lost contact with each other and the basal lamina were rearranged initially in clumps, but were later arrayed along tracheae within a reformed basal lamina. In cultures of disaggregated fat body cells, haemocytes pulled cells together to form new aggregates. A similar role for haemocytes in tissue reorganization during metamorphosis was proposed. Some authors have interpreted the changed relationships between fat body cells, and their content of autophagic vacuoles, as evidence for cell death and disintegration of the larval tissues. In *Leptinotarsa,* Labour, G. observed that extensive autophagy was not uniform throughout the fat body, and he concluded that only some parts of the fat body undergo autophagy and survive in the adult, whereas other areas disappear completely. However, Krishnakumaran, A. and co-workers recovered intact pieces of larval fat body, transplanted into silkmoth pupae, from the newly emerged adults. This result supports the ultrastructural evidence for the persistence of most larval cells into the adult. What remains to be clarified is the generality of cell death or its absence during metamorphosis of larval tissue.

Cell Remodelling

The transformation of larval fat body into the adult tissue begins during the last larval stadium, with the precisely executed autophagy of cell organelles. In *Calpodes* the process involves the isolation and lysis of peroxisomes, mitochondria and RER, in that order, with the RER being cut off in fragments of about mitochondrial size. In a morphometric study, Larsen, S. showed that the cells remain depleted of the lysed organelles throughout most of the pupal stage. Regeneration of RER and peroxisomes begins in the pharate adult and continues after emergence. Mitochondria increase in number following a phase of division 22-34 h after emergence. The new individuals at pupal ecdysis. Although preparation for diapause takes longer than preparation for pupation, both follow the same pattern. They are characterized by the loss of RER and mitochondria by autophagy and the accumulation of protein granules. Since pupation can be prevented and diapause induced experimentally by a single application of JH to the larva, it appears that in this species JH permits, or does not inhibit, the formation of protein granules.

TABLE 6.5: HORMONAL CONTROL OF HETEROPHAGY, PROTEIN STORAGE, AUTOPHAGY AND HISTOLYSIS

Order and species	*Method*	*Hormone*	*Effect*
Diptera			
Calliphora erythrocephala	EM,TR	20-OH Ec	Increased acid phosphatase
Calliphora vicina	EM	20-OH Ec	Autophagic vacuoles formed. (L)*Autophagic vacuoles formed. (A)
Drosophila melanogaster	LM, Tr, Vt	20-OH Ec	Protein storage, lysosomal granules Protein storage. (L)
	LM	+JH	Histolysis of larval fat body in adult. (L)
Lepidoptera			
Bombyx mori	EM	20-OH Ec, (-JH)	Protein storage granules, autophagic vacuoles formed. (L)
Calpodes ethlius	LM, EM, Vt	20-OH Ec	Protein storage granules, autophagic vacuoles formed.
	Cm		Acid protease increased. (L)
Ephestia kuhniella	EM	20-OH Ec	Protein storage granules formed. (L)
Mamestra brassicae	EM, Vt	20-OH Ec, (- JH)	Protein storage granules, autophagic vacuoles formed. (L)

Control of Autophagy and Histolysis

The precisely timed beginning of autophagy during metamorphosis and cell remodelling has been correlated with changes in the haemolymph ecdysteroid titre. The isolation and lysis of peroxisomes and mitochondria coincide with increased hormone levels in *Calpodes*. Dean, R. has shown that the presence of 20-hydroxyecdysone is a sufficient stimulus for induction of autophagy in fat body of fifth-stage *Calpodes* larvae *in vitro;* there is no autophagy in hormone-free medium. However, autophagy once induced does not require the continuing presence of the hormone. This shows that fat body cells are intrinsically capable of sustaining autophagic activity. Why does autophagy not occur in response to the elevated ecdysteroid titre which precedes each larval-larval moult? In penultimate-stage larvae of *Mamestra,* ecdysteroids induce fat body autophagy when injected during the last day, but not during the preceding 3 days of the stadium. A reduction in the JH titre occurs during the penultimate larval stadium so that JH-inhibition of ecdysteroidinduced autophagy might be invoked to explain the result. However, when these authors incubated fat body of different ages in the presence of ecdysone, but no JH, only cells taken from larvae on the day preceding fourthto fifth-instar ecdysis underwent autophagy. It was suggested that the cells require a period of time to gain competence to respond to ecdysteroid after exposure to high JH levels. The change in competence of the fat body may be similar to the change in commitment of the epidermis in response to the commitment peak of ecdysteroid in *Manduca* and *Calpodes.*

The control of fat body cell lysis in *Drosophila* is less well understood. Of the larval cells which survive metamorphosis in wild-type flies, about 95% disappear by the end of the second day of adult life. Surviving larval cells in flies homozygous for the ap^4 mutant fare better. In ap^4 flies the larval cells disappear more slowly during the first 2 days of adult life and their number remains unchanged over the next 3 days. Transplantation experiments with fat body from newly ecdysed adults showed that ap^4 cells survive better because their environment induces less histolysis and not because of any intrinsic cellular property. Also, in both wild-type and ap^4 flies the internal environment becomes decreasingly potent in inducing lysis with increasing age. There are anterior-posterior gradients of heterophagic activity and cell separation in *Drosophila,* and regional differences in acid phosphatase activity in *Calliphora erythrocephala.* These observations support the view that fat body cells intrinsically control their ability to respond to the milieu that triggers granule formation and autolysis.

In an earlier study Butterworth, F. and Bodenstein, D. performed transplantation experiments to discover the fate of late second-instar fat body cells in the adult environment. These cells showed no histolysis in normal adult males but when implanted together with ring glands from late third-instar larvae massive degeneration was inferred because the cells of the transplant could not be recognized. When second-instar cells were implanted into recently emerged females, cytological changes were also induced but were less complete and later than those seen in males with ring gland implants. Implants into ovariectomized females showed no degeneration, whereas those in males containing a transplanted ovary showed some histolysis. Since higher dipteran larval ring glands secrete ecdysone and so do dipteran ovaries, these results can be explained by ecdysteroid induced autophagy of fat body cells. Butterworth, F. later reported that fat body cells from mid and late third-instar larvae transplanted into adult males did eventually begin to disappear, but not nearly as rapidly as the cells of the host which had undergone metamorphosis. He suggested that rapid lysis must depend on additional preparative changes which make the cells responsive to the lytic environment in the adult. These studies are all complicated by the difficulty of quantifying cell degeneration and death discussed, for example, by Butterworth, F.. 20-Hydroxyecdysone is also implicated in the induction of autophagy in *Calliphora.* Precocious autophagy and storage granule formation is induced in the fat body of younger larvae when transplanted into older larvae and by 20-hydroxyecdysone injection into premetamorphic larva. The problem is complicated by the observation of Day, M. who found that ring gland extirpation prevented the histolysis of larval fat body in adult *Lucilia,* and concluded that the corpora allata are required. Postlethwait, J. and Jones, G. concluded that the ring gland factor is JH, since fat body histolysis is inhibited in isolated abdomens of *Drosophila* but can be stimulated by the application of a JH analogue. Perhaps a less equivocal result on the factors inducing autophagy in dipteran fat body will be found through *in vitro* studies where direct hormone action can be observed.

The respective roles of ecdysteroids and JH are not understood. In *Calliphora* there is increased activity of lysosomal enzymes in fat body at metamorphosis and this increase can be induced by 20-hydroxyecdysone. In *Calpodes ethlius* increased activity of acid hydrolases is correlated with the increased haemolymph ecdysteroid titre. These increases in lytic enzymes do not by themselves ensure degradative activity, for in *Calpodes* and other insects, destruction of organelles in autophagic vacuoles occurs side by side with the storage of haemolymph proteins in granules. Hydrolase activity remains relatively high thoughout the puparial stage of *Drosophila melanogaster* without affecting the granules and through the pupal stage of *Calpodes*.

It has been suggested that RNA synthesis 9-12 h before increased hormone titre is a prerequisite for induction of autophagy by 20-hydroxyecdysone in *Mamestra*. Similarly, Postlethwait, J. and Jones, G. have proposed that protein synthesis is necessary before J11-stimulated histolysis of larval fat body of *Drosophila will* occur. The nature of such protein(s) is unknown.

BIOCHEMISTRY OF THE FAT BODY

The fat body is the principal tissue for intermediary metabolism in insects. While appearing superficially to be an adipose storage tissue, the fat body is in reality a tissue of considerable metabolic activity and is the main source for the hemolymph proteins, lipids and carbohydrates that serve as precursors for metabolism in other tissues. The fat body undergoes growth and development along with the other insect tissues and its functions change in accordance with the developmental stage of the insect. Because of its changing metabolic role and its integral position in maintaining metabolic homeostasis, the fat body is a target tissue for hormones and serves with increasing frequency as a model for examining endocrine regulations of metabolism.

The fat body has a broad metabolic role and the present review will attempt to discuss those areas that are receiving frequent research attention. Although this review seeks to be comprehensive, it concerns itself primarily with studies published during the nearly two decades that have elapsed since the last general review of fat body biochemistry. The present discussion will focus primarily on unique fat body functions and their contribution to the overall physiology of the insect, and will not emphasize general metabolic processes such as glycolysis, the TCA cycle, oxidation of fatty acids, etc. that are common to most animals.

It was deemed essential to be selective for the topics and articles for this review. Undoubtedly some items of interest were omitted because they were overlooked or their involvement with the fat body was not clearly evident. In the latter case, examples are studies that use whole-body homogenates for which the results can be attributed mainly to the fat body. The quantity of literature dealing specifically with the fat body makes unnecessary the inclusion of studies such as these, that use imprecise methods.

The subject-matter of this chapter focuses mainly on the biochemistry of the fat body with only minimal references to structure since an entire chapter in this series reviews fat body structure. Nevertheless, structural organization and biochemical function are complementary so that some redundancy is necessary and inevitable.

FAT BODY ORGANIZATION

The insect fat body is structurally heterogeneous. The fat body arrangement is described variously by Buys, K. as: pads, layers, sheets, masses, bladders, ribbons or bead-like. Its colour is usually white, or either pale yellow or green. It is generally the most obvious organ in the insect abdomen where the largest deposits of fat body are located. However, in addition to the large abdominal deposits the fat body is also scattered throughout the remainder of the body where it surrounds other organs including the brain, central nervous system, gut, muscles and gonads. This diffuse distribution is probably an adaptation to the open hemocoel. Metabolites are transported to insect tissues, in part, by diffusion along hemolymph concentration gradients. By being widely distributed, the fat body reduces the diffusion distances between itself, as the source tissue for the metabolites, and the peripheral tissues where the metabolites are used.

The fat body exhibits regional differentiation. Peripheral or subcuticular fat body is found near the body wall outside the musculature, and central or visceral fat body surrounds the alimentary canal. The two regions may have different functions. The visceral tissue appears more biosynthetically active while the peripheral tissue stains more heavily for glycogen and appears to have a storage role. The two areas of fat body are easily distinguished in *Blaberus discoidalis* cockroaches since subcuticular fat body lacks the symbiote-containing cells (mycetocytes) that are common in the lobes of the visceral fat body.

Regional biochemical differences are demonstrated by localized deposits of eye pigments in the fat body of *Drosophila melanogaster* fruit-flies. Kynurenine deposits form in the anterior fat body of larvae but not in the posterior tissue. During adult metamorphosis, these pigmentcontaining cells localize in the future head-thorax region. In the *Ore-R* strain of *D. melanogaster* only the anterior fat body makes kynurenine, but tryptophan feeding induces tryptophan pyrrolaseactivity in other parts of the tissue so that other cells also begin to produce kynurenine. The number of responding cells is related to the amount and duration of tryptophan feeding and to larval age. These results indicate the presence of a tryptophan pyrrolase gene that is inactive in most fat body cells but is inducible at differing levels of sensitivity by the presence of the substrate. In a related situation, wild-type *D. melanogaster* show localization of the eye pigment isoxanthopterin in only the posterior fat body. The regional formation of isoxanthopterin results from substrate localization since the biosynthetic enzyme for isoxanthopterin is ubiquitous throughout the tissue. Further regional differentiation is observed in the fat body of D. *melanogaster* larvae where posterior cells store greater concentrations of protein granules than do anterior cells. Finally, in the bloodworm, *Chironomus thummi,* hemoglobin is synthesized by subcuticular fat body, whereas the central tissue appears to function for metabolite storage.

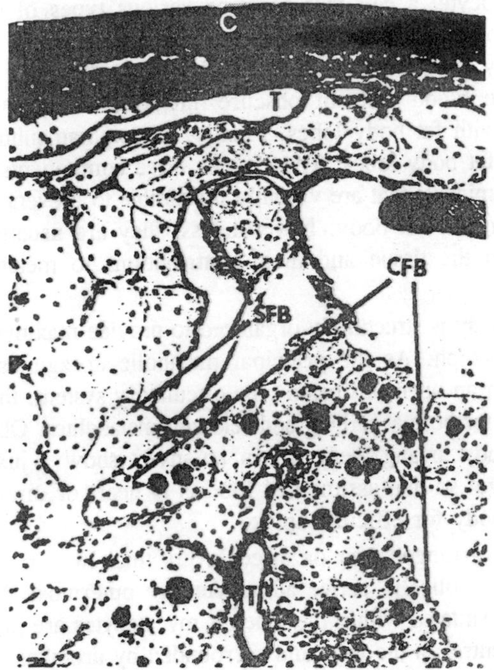

Fig. 7.1. Localization of mycetocytes within the central fat body relative to the subcuticular fat body in 4-day-old adult male.

The predominant fat body cell is the adipocyte. The older literature uses the term "trophocyte" which seems to this author to be a misnomer. Trophocyte suggests a "nourishment or food" cell. The lipoidal appearance of the fat body may suggest a nutrient-storage function, but we know now that the fat body is more than a storage depot. Both trophocyte and adipocyte are used in the current literature. Therefore, in keeping with the vertebrate terminology for cells from fatty tissues, I propose that the term adipocyte be used in the future in place of the inexact "trophocyte" terminology.

The fat body is composed of either two or three cell types. The predominant metabolic-storage cells are the adipocytes which resemble vertebrate brown fat in that they contain multiple fat droplets rather than the single droplet found in vertebrate white fat. Furthermore, filamentous mitochondria, free ribosomes and glycogen are frequent in adipocytes of both the fat body and brown fat. The second common cell type is the urocyte that sequesters uric acid for storage-excretion. A third cell type that is found in the fat body of some insect species is the mycetocyte; a cell that contains various types of symbiotes, especially bacteria-like organisms. Hemocytes are scattered coincidentally between the lobes of the fat body, and oenocytes, adipocyte-like blood cells of obscure function, are often found in conjunction with fat body lobes. Finally, Malpighian tubules ramify between the fat body lobes. These latter three components constitute cellular contaminants that are virtually impossible to prevent or remove when isolating the fat body. Nevertheless, they are usually a minor component in the tissue and their contributions to metabolism are negligible.

The fat body is structurally organized to provide maximal exposure to the hemolymph. As the principal metabolic-storage tissue in an organism with an open, diffusion-type circulatory system, the fat body is well-suited for both absorbing and releasing metabolites. Observations on the fat body arrangement within an insect show a juxtaposition with the midgut so that assimilated food is in near, or direct, contact with the fat body for efficient uptake.

Cells are arranged two or three layers thick within a fat body lobe. The metabolically active adipocytes are outermost in the lobe with urocytes situated within the lobe. If mycetocytes are present they are usually central in the lobe and surrounded by urocytes. The more exterior position of the adipocytes gives them maximal exposure to the hemolymph for metabolite exchange. Urocytes, by contrast, are degenerate cells that lack cytoplasmic structures such as ribosomes, mitochondria or endoplasmic reticulum. Presumably, urocytes have low metabolic activity and are mainly storageexcretion depots for uric acid, a nitrogenous end product in many insect species or a storage form for nitrogen. The spatial arrangement between urocytes and mycetocytes suggests that they interact. It is reported that bacteroids convert uric acid to either TCA cycle intermediates or to nitrogenous products used for growth.

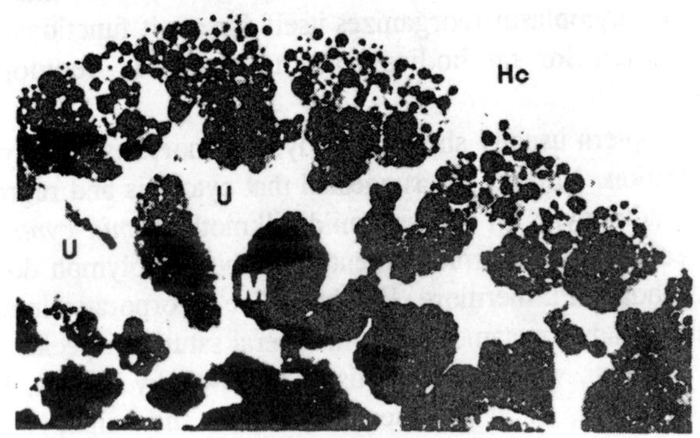

Fig. 7.2. Distribution of cell types in a typical lobe of the fat body from adult male.

In summary, it is obvious that the structure of the fat body is highly adapted to the unique organization and physiology of the insect. Whereas the fat body appears initially as a diffuse and amorphous storage tissue, closer scrutiny suggests its amorphic distribution throughout the insect body is really an adaptation to the open circulatory-diffusion system. Furthermore, the fat body has regions of specialization. The tissue located in the center of the body, nearest the intestine, appears most metabolically active, while the subcuticular fat body may have a greater storage function. The individual fat body lobes are organized so that metabolic cells are assured maximum exposure to the circulatory system, while cells involved in storage and the recovery of storage products are more secluded within the lobe.

METAMORPHOSIS

There are two aspects for consideration in fat body metamorphosis. The first concerns structural reorganization of the tissue while the second concerns biochemical changes. The biochemical changes are discussed throughout the remainder of this chapter as aspects of metabolism rather than in this section as aspects of metamorphosis.

The fat body undergoes three types of imaginal metamorphosis. In the first type, adult adipocytes are derived by structural rearrangement of the cytoplasm within existing larval adipocytes. In the second type, adult fat body forms by lysis of larval adipocytes followed by replacement with precursor imaginal cells. Finally, the fat body of the Colourado potato beetle, *Leptinotarsa decemlineata* forms from a combination of the first two types.

In some insect species, larval adipocytes persist into the adult where their cytoplasm reorganizes itself for adult functions. This has been reported for fat bodies from mosquitoes, Lepidoptera and cockroaches.

Lepidoptera usually show fat body metamorphosis by cytoplasmic reorganization. Ishizaki, H. suggested that cytolysis and regrowth may occur in adipocytes of the saturniid silkmoth, *Samia cynthia ricini,* with release of cytoplasmic contents into the hemolymph during adult metamorphosis. Furthermore, [^{3}H]thymidine incorporates into fat body DNA during adult metamorphosis in several saturniid species, including *S. cynthia ricini,* which suggests tissue regrowth by mitosis. However, Krishnakumaran, *A. et al.* question whether the adipocytes actually disintegrate as suggested by Ishizaki. They find, by histology, that the adipocytes separate and discharge their contents, hence the fat body appears "mushy" during metamorphosis. But in fact, the adipocyte cell membranes remain intact and the tissue later reorganizes into a definitive structure. Dissociation of the adipocytes appears ecdysteroid regulated and occurs *in vitro* in a dose-dependent relationship with 20-hydroxyecdysone (20-OH ecdysone).

A definitive experiment was performed with the lepidopteran fat body to confirm whether the tissue persisted and reorganized or histolyzed and regrew. Larval fat body DNA was labeled with [^{3}H]thymidine in the silkmoth *Hyalophora cecropia.* After pupation, a piece of prelabeled tissue was transplanted into an unlabeled early pharate adult.

The intact, labeled fat body transplant was located later in the resulting adult and showed adult characters. These studies confirm that the larval adipocytes of Lepidoptera persist through metamorphosis and differentiate into adult adipocytes.

Cytolysis and regrowth is illustrated clearly in higher Diptera. The larval fat body disintegrates in an anterior to posterior direction shortly after pupariation in the cyclorrhaphous Diptera. Whitten, J. reported that disintegration of the fat body results from loss of the surrounding basement membrane, but that although the adipocytes are free, they remain *in situ*. The freed larval adipocytes constitute the "pupal" fat body. Most of the pupal adipocytes undergo cytolysis about the third day of pupariation; however, some persist into the adult where they eventually undergo cytolysis and disappear. Adult adipocytes arise either from imaginal precursor cells located under the pupal integument or from embryonic cells in the posterior of the larva. Hemocytes are also proposed as precursors for imaginal adipocytes.

The imaginal environment causes histolysis of those larval adipocytes that persist into adults. Approximately 1000 larval adipocytes survive into the adult in *Drosophila melanogaster*. Normally, 90% of these cells disappear within 2 days of adult emergence and are totally gone by 4 days. However, larval adipocytes survive for longer times when transplanted into older flies. If the larval cells are pre-exposed to the environment of young flies for 1 day before transplantation into older flies, then the larval cells degenerate as quickly as if kept continuously in the lytic environment of young flies.

Endocrine factors appear to regulate the lytic effect. For example, larval adipocytes degenerate rapidly in young *Musca domestica* houseflies when vitellogenic ovaries are present, but either ovariectomy or allatectomy of the young flies causes persistence of the larval adipocytes. Furthermore, treatments with juvenile hormone (JH) analog increase by 3-fold the cytolysis rate of persistent larval adipocytes in abdomens isolated from adult *D. melanogaster*. Cycloheximide administration blocks the lytic action of JH on larval adipocytes and indicates that protein synthesis facilitates the lytic process. Histolysis of larval adipocytes occurs only after protein body formation and in the presence of either the ring gland, corpora cardiaca-allata complex or functioning ovaries.

In combination with our present knowledge of the hormone balance in adult Diptera and several past references to fat body histolysis, it can be suggested that ecdysteroids stimulate the lysis of larval adipocytes by the formation of lysosomal protein granules. Acid phosphatase is present in protein granules of *Calliphora erythrocephala* as shown by dissociation of the granules at pH > 8.6. Furthermore, both acid phosphatase activity and protein granules appear in the fat body of *C. erythrocephala* larvae after treatment with 20-OH ecdysone. Hence, these adipocyte protein granules are lysosomal in origin and function, and they form in response to increasing ecdysteroid titer at the onset of metamorphosis. It is now known that ovaries secrete ecdysone in adult mosquitoes. Therefore, the larval ring gland and the active adult ovary may be sources for ecdysone so that the presence of either structure in an adult fly makes the environment lytic for persistent larval adipocytes. It is not clear why allatectomy causes persistence of larval adipocytes or why JH treatment induces the lytic milieu. Possibly, in the adult fly, JH acts in conjunction with a brain neurohormone to stimulate the ovaries to secrete ecdysone similar to the case in mosquitoes.

A third type of fat body metamorphosis combines both cytolysis-regrowth and cytoplasmic restructuring. Some parts of the pupal fat body undergo lysis in *Lcptinotarsa decemlineata* while other pupal adipocytes persist into the adult and show cytoplasmic reorganization.

FAT BODY BIOCHEMISTRY AND ITS REGULATION

Basal Metabolism

Basal metabolism is measured by the rate of oxygen consumption and is a function of the rate of energy production which reflects the total biosynthetic level in the tissue. Protein synthesis is the most energy-demanding process in the cell so that oxygen use is mainly a function of the rate for protein synthesis.

Oxidative metabolism occurs in the fat body of insects via the TCA cycle and electron transport just as it does in other animals. However, it should be noted that respiratory enzyme activities often fluctuate both with the developmental stage of the insect and within any given instar. For example, fat body cytochrome c oxidase activity is high at the start of the fourth nymphal instar in *Schistocerca gregaria,* declines by 97% by the end of the instar, then increases sharply at the adult molt and is high in young adults. There is now little reason to question the integrity of the complete complement of usual mitochondrial enzymes in the insect fat body and with activities and substrate preferences comparable to those reported for rat liver mitochondria.

The fat body is a tissue rich in lipid stores, but carbohydrate is most frequently its main energy source. A respiratory quotient (RQ) of 0.94 is determined for adult fat bodies from *Leucophaea maderae* cockroaches. However, in adult male *S. gregaria* the fat body has an RQ of 0.64-0.72, indicating that lipid is the main substrate for energy in this insect. Addition of a corpora cardiaca extract to *in vitro* fat bodies from adult female *L. maderae* increases the fat body O_2

consumption. This increase in O_2 uptake occurs because of a 30-35% increase in the rate of oxidation of lipid instead of carbohydrate. The shift from the use of carbohydrate to lipid for energy production is a compensation for the increase in rate of conversion of carbohydrate substrates to hemolymph trehalose that occurs in response to the presence of a hypertrehalosemic factor in the corpora cardiaca extracts.

It is expected that hormones affect the basal metabolic rate of the fat body since it is the principal biosynthetic tissue of the insect and should respond to metabolic demands associated with endocrine regulations of growth and development. Luscher, M. reported a 50% decrease in fat body respiration, a 70% decrease in fat body protein synthesis and arrested oocyte growth after decapitation of adult female *Nauphoeta cinerea* cockroaches. Corpora allata implants reverse these effects, suggesting that JH is the main hormone regulating all of these processes. The effect of JH on fat body respiration in female insects is related to the JH-dependent production of yolk precursors by the fat body for vitellogenesis. In addition to the JH effect, a small but significant increase in respiration also occurs after brain-corpora cardiaca implants, suggesting some additional neuroendocrine stimulation to the basal metabolism. Corpora cardiaca extracts stimulate respiration by *in vitro* fat bodies from adult female *L. maderae* if the initial fat body respiration is low, but are inhibitory if the initial respiration is high. Furthermore, fat body respiration is reduced by about 40% in adult male *Blaberus discoidalis* 30 days after extirpation of the corpora cardiaca-corpora allata complex. This reduction in fat body basal metabolism is corpora cardiacarelated, and the effect is manifested in fat body mitochondria at the level of electron transport activity. This latter result suggests neuroendocrine regulation for the energy-generating capacity in fat body mitochondria.

Therefore, fat body respiration responds to two endocrine regulations. The corpora allata effect fat body respiration associated with protein synthesis for ovarian development. In addition, neuroendocrine factors increase fat body respiration by affecting the types of substrates used and the intrinsic capacity of mitochondria for electron transport. The corpora allata effect accelerates basal metabolism through respiratory control by ADP generated from endergonic, biosynthetic processes, while the neuroendocrine effect determines the exergonic capability within the adipocytes.

New mitochondria form and show synchronous development in imaginal adipocytes shortly after adult emergence. This mitochondriogenesis raises the exergonic capability of the fat body by increasing its number of functional mitochondrial units. Adult adipocytes form in *Calpodes ethlius* by cytoplasmic reorganization of larval adipocytes, and larval mitochondria are destroyed before pupation by engulfment into autophagic vacuoles during the cytoplasmic degeneration associated with metamorphosis. Pupal adipocytes contain few mitochondria, but new mitochondria form in the young adult. During the first 22 h of adult life, mitochondrial remnants from the pupal adipocytes increase in volume. From 22 to 34 h the mitochondria decrease in volume but increase in number. During this latter period the inner membrane frequently transects the mitochondria into two inner compartments, which suggests mitochondrial division. These observations indicate that new adipocyte mitochondria form in adults by enlargement and division of existing mitochondria carried over from the larva through the pupal stage. In the cockroach *B. discoidalis,* adipocyte reorganization occurs during the first 4 days of adult life, and fat body mitochondria increase their respiratory capacity by 3-fold during the first 10 days of adult life. The respiratory increase is associated with elevated heme synthesis between 4 and 6 days of adult age and an increase in mitochondrial cytochromes. The increase in heme synthesis is dependent on a "cytochromogenic hormone" from the brain-corpora cardiaca complex.

Mitochondriogenesis appears to be an integral aspect of adipocyte metamorphosis. Presumably the fat body converts from a mainly nutrient storage function in immature insects into a more biosynthetically active adult tissue. This is especially true for the adult female fat body which must produce large quantities of protein and lipids for transfer to ovaries for vitellogenesis. The maturation of new adipocyte mitochondria insures an adequate exergonic capacity to fulfill the increased endergonic demands of vitellogenesis, and appears neuroendocrine regulated at the level of the cytochromes.

Nucleic Acids

DNA

Fat body DNA can either remain constant or vary depending on the insect species and its developmental stage. Although adipocyte mitosis is observed in the *C. ethlius* fat body at the end of the penultimate larval instar, DNA synthesis occurs without further mitosis during the last larval instar. This increase in adipocyte DNA without subsequent cell division is indicative of polyploidy and endomitosis. Incorporation of [^{3}H]thymidine shows evidence for DNA synthesis during larval life in the fat body of saturniid silkmoths. However, in *Calliphora erythrocephala,* the fat body DNA remains constant at about 0.016 pg DNA per fat body throughout larval life. In *Hyalophora cecropia* the fat body DNA content did not change during adult metamorphosis, although the DNA content doubled in the wing epidermis during this time. In a comprehensive study of saturniid silkmoths, Krishnakumaran, A. *et al.* reported that fat body DNA is synthesized during adult metamorphosis in some species but not at the time of the larval-pupal transformation.

Hormones may influence the ability of adipocytes to synthesize DNA at various stages of growth. Exposure to JH induces DNA synthesis in pupal adipocytes of saturniid silkmoths during a stationary pupal-pupal molt if the cells are also exposed to ecdysteroids. However, adipocytes do not synthesize DNA in pupal abdomens grafted to adult male silkmoth abdomens where the endogenous JH titer is naturally high and ecdysteroids are absent. This suggests that the ecdysteroids are the hormones that facilitate DNA synthesis in the pupal fat body. In adult insects it is generally accepted that fat body DNA does not increase. However, in young adult *Locusta migratoria,* adipocyte DNA doubles during the first previtellogenic period. This doubling of adipocyte DNA in adult locusts is presumably an early maturational response by the fat body to JH, since this increase in DNA content does not occur after allatectomy. DNA synthesis may be needed for fat body maturation preparatory to the later production of yolk constituents. Polyploidy in adipocytes of adult female locusts may enhance the capacity of the adipocytes to produce yolk precursors since the increase in DNA copies and their accompanying genes permit the production of increased amounts of biosynthetic enzymes. These findings suggest that the hormones that regulate adipocyte DNA synthesis change between pupal and adult life as the tissue switches from growth to reproductive functions.

DNA was characterized from adipocytes of diapausing pupae and pharate adults of *H. cecropia* by isopycnic ultracentrifugation. The DNA patterns remained similar regardless of the developmental state of the tissue source. A major DNA band occurred at an equilibrium density of 1.693 with two minor satellite DNA bands comprising 6% and 3% of the DNA and located at densities of 1.705 and 1.709, respectively. These densities indicate G + C amounts of 34% for the major band and 46% and 50%, respectively, for the 1.705 and 1.709 bands. Renaturation kinetics showed 30% redundant sequences in the genome and 70% unique sequences.

Fat body DNA increases in pharate pupae of the silkworm *Bombyx mori,* but not by new DNA synthesis. Administration of [³H]thymidine fails to label accumulating adipocyte DNA and suggests that the new DNA originates by degradation and transport to the fat body from another tissue. The silk glands degenerate at this time in pharate *B. mori* pupae and their DNA disappears. This suggests the silk glands are the source for the accumulating fat body DNA. Analysis of the pupal adipocyte DNA in *Bombyx mori* separates three types: α, β_1 and β_2. All three types have a G + C content of 38% although the Lx-DNA is several times larger in molecular weight than the β-DNA. No β-DNA occurs in larval adipocytes, but it predominates in pupal cells. Tissues that degenerate, such as the pupal silk gland or midgut, contain β-DNA; whereas the larval silk glands and developing tissues, such as the integument, muscles and gonads, contain only (Y-DNA. Although β-DNA increases in adipocytes during early pupal life, only α-DNA becomes labeled by [³H]thymidine. This indicates that the fat body β-DNA has already formed before its appearance in the adipocytes. Comparisons of DNA labeling times and specific activities between various tissues indicate further that the β-DNA found in the pupal fat body forms from the degradation and transfer of larval silk gland DNA. The β-DNA is located in the cytoplasm of the adipocytes and is ultimately transferred to the ovaries during vitellogenesis for use as a source of nucleic acids for embryogenesis.

RNA

Fat body RNA/DNA ratios provide a relative measure of protein synthetic capacities but not necessarily of synthetic activities. Price, G. reported exceedingly high ratios of 60 to 125 for RNA/DNA in fat bodies of C. *erythrocephala* larvae. These ratios are more than 50 times greater than those generally reported for vertebrate tissues. However, other reports on fat body RNA/DNA ratios are not so extreme. The fat body of adult male *Blaberus discoidalis* shows an increase in RNA/ DNA from a value of 3.5 at adult emergence to a value of 7 by day 6. RNA/DNA ratios in the fat body of adult *Leptinotarsa decemlineata* range from 6 to 15 for females but are only about 4.0 for males. In contrast to the adult situation, fat body RNA/ DNA ratios decrease from a high of 8 in early fifth-instar larvae of *Philosamia cynthia ricini* to values of 2 to 3 at pupation due to the loss of rRNA from the microsomal fraction.

Developmental changes in fat body RNA are illustrated in *Calliphora erythrocephala*. Adipocytes have the highest RNA content during early larval life with a decline as pupariation approaches.

Examination of the fine structure of the adipocytes shows that ribosomes are abundant in young larvae but disappear before pupariation. Protein synthesis follows this same pattern. Measurements of RNA content per fat body at the time of pupariation suggest that the RNA content is still high even though the ribosomes are degraded. As much as half of the larval RNA is fat body RNA, and most of the adipocyte RNA in young larvae is ribosomal with as much as 25 mg of ribosomes per gram of fat body. This amount of ribosomes is eight times greater than that found in vertebrate liver. During the last 3 days of larval life, 85-90% of these ribosomes are destroyed in association with an increase in degraded 4S-7S, insoluble RNA. The degraded adipocyte RNA is not replaced and forms an inert, insoluble "pellet RNA" after centrifugation at 20,000 g for 10 min. This inert RNA is formed exclusively from ribosomes and is excreted in the meconia after adult emergence. Pharate adult cells synthesize 90% of their ribosomal RNA anew during metamorphosis.

In contrast to Diptera, little RNA synthesis occurs in the fat body of Lepidoptera at the onset of adult metamorphosis. A general survey of RNA synthesis in the tissues of several silkmoth species shows that some fat body RNA synthesis occurs during the early half of pharate adult development but declines during the last half. No measurable differences occur in fat body RNA/DNA ratios between diapause termination and the onset of metamorphosis in *Hyalophora cecropia* pupae; however, fat body RNA synthesis increases by 3fold during the first few days of metamorphosis, then ceases by day 3. In *Antheraea pernyi,* fat body rRNA forms before the onset of visible adult development and is one of the earliest events of diapause termination. However, after the third day no further rRNA accumulates in the fat body of the pharate adult even though [^{14}C]uridine incorporation still occurs. This suggests that the rate of rRNA synthesis is approximately equal to its rate of breakdown so that no net increase in fat body rRNA occurs at this time. These studies do not identify whether new mRNA is produced at the onset of metamorphosis or whether pre-existing mRNA transcripts survive dormancy and are translated on newly formed ribosomes. The status of fat body RNA and its synthesis in response to hormones at diapause termination may be a useful model for elucidating fundamental endocrine effects on cell metabolism and differentiation. The dormant adipocyte is at a minimal metabolic level, and primary endocrine effects should be observable relative to a minimum of "background" metabolism.

RNA Polymerase and Hormone Effects

RNA formation is an early fat body response after exposure to either ecdysteroids or JH. Either hormone stimulates RNA synthesis in adipocyte nuclei isolated from C. *erythrocephala* larvae, but the simultaneous presence of both hormones inhibits their individualeffects. Hybridization of adipocyte RNA and DNA increases after exposure of the adipocytes to either hormone; however, DNA hybridization to RNA produced by either one of the hormones is blocked if the DNA is pretreated with RNA produced in response to the other hormone. This suggests that both hormones regulate common genes in the fat body but are antagonistic for the regulations when present together.

Ecdysteroids, in particular, stimulate RNA synthesis in adipocytes. In *A. pernyi* larvae, fat body RNA synthesis increases by 9-fold within 90-120 min after ecdysone injection and returns to its original base level by 270 min. Fat body RNA synthesis increases by over 5-fold, 8 h after 20-hydroxyecdysone (20-OHecdysone) treatment of adult female *Gromphadorhina portenosa* cockroaches along with a concomitant 3-fold increase in protein synthesis. Injections of 20-OH-ecdysone stimulate the production of poly(A) RNA in adipocytes of *Calliphora vicina* larvae which indicates hormone induction of new messenger RNA transcripts. Both ecdysone and 20-OH-ecdysone stimulate fat body RNA synthesis with 20-OHecdysone having the greater effect, but only when endogenous ecdysteroid titers are low. Presumably, hormone treatment when endogenous ecdysteroid titers are high fails to stimulate RNA synthesis above its already high base level. DNA-RNA hybrid competition studies show that RNA from untreated control larvae is not competitive with RNA from hormone-treated larvae, and indicates that new types of RNA appear in response to the ecdysteroids. Both mRNA and rRNA increase in response to ecdysteroid treatment with the majority being rRNA. Similar studies with larvae of *Sarcophaga peregrina* indicate that injections of 20-OH-ecdysone preferentially activate RNA polymerase I in adipocyte nuclei for rRNA synthesis. These results indicate that, although part of the effect of ecdysteroids is on expression of new genomic information, a major response is to make the adipocyte competent for increased protein synthesis by enlarging its ribosomal pool.

There is considerable interest in the RNA polymerases of insect tissues because of their possible involvement in endocrine regulation of growth and metamorphosis. Activation of RNA polymerase may be an essential component in hormonal regulation of specific genes for cell growth and differentiation. Ecdysteroids and JH both appear to act on transcription, hence RNA polymerase activities may be significantly influenced by these hormones. The majority of fat body RNA polymerase is sensitive to the inhibitor αamanitin during larval life in the tobacco hornworm, *Manduca sexta,* which suggests that the predominant larval enzyme is RNA polymerase II for mRNA production. At the time of wandering, when ecdysteroids appear, RNA polymerase activity declines in association with a change in the enzyme from a soluble to an insoluble form. The insoluble RNA polymerase continues into adult life, but even though the enzyme does not disappear, it is doubtful that it is still functional for gene transcription. The conversion from soluble to insoluble RNA polymerase coincident with the burst of ecdysteroid titer that occurs at the onset of wandering suggests that the structural form of RNA polymerase is endocrine regulated. Sridhara, S. and Gilbert, L. isolated RNA polymerase from the fat body and found two bands sensitive to a-amanitin and stimulated by Mn^{21}. They suggest that the two bands represent multiple forms of the enzyme, and speculate that the major band is the free enzyme and the minor band a bound form. Fat body RNA polymerases I, II and III are also described for larvae of the corn earworm, *Heliothis zea.* Characterization by DEAE-Sephadex chromatography of the *H. zea* fat body RNA polymerases resolves the usual three major bands but also shows two additional minor bands with RNA polymerase activity.

Most of the studies to date have been inconclusive with respect to relationships between fat body RNA polymerases, hormones and regulations of growth and metamorphosis. However, as the fat body enzymes become better characterized and an optimal assay system developed, the fat body could prove valuable as a model for elucidating how hormones regulate genetic expression. This is particularly true for fat body proteins, such as vitellogenin, which are well-characterized and produced in large amounts at specific times in the life cycle in response to hormone stimulation.

Proteins and Protein Synthesis

The fat body is the principal synthetic source for hemolymph proteins. Shigematsu, H. first demonstrated that soluble proteins released *in vitro* from the fat body of *Bombyx mori* are electrophoretically similar to hemolymph proteins, and the rate of fat body protein secretion *in vitro* matches the rate of hemolymph protein formation *in vivo.* The fat body produces many unique and physiologically significant proteins such as vitellogenins for oocyte maturation; stage-specific, amino acid-storage proteins such as calliphorin, drosophilin and manducin; diapause proteins and even hemoglobin in the chironomid bloodworms. The fat body is the source for lipid-binding proteins, and the JH-binding protein that protects circulating JH from degradation. Conversely, the fat body produces the specific esterase that degrades JH. Finally, the fat body has a unique role in protein storage and conservation by forming large proteinaceous storage granules that degrade and serve as a source for proteins and amino acids during metamorphosis. Various aspects of these proteins and their relationships to the fat body were reviewed by Chen, P.; Price, G.; Wyatt, G. and Pan, M.; and Keeley, L..

Endocrine Effects

Neurohormones appear to increase the general protein synthetic capacity in the fat body. The neuroendocrine effects are especially apparent in female adult insects during yolk precursor production. In female adult insects the vitellogenic proteins comprise a major portion of the proteins synthesized by the fat body and their formation is JH-regulated. Nevertheless, the corpora cardiaca also affect fat body protein synthesis in female adult insects. The rate of protein synthesis increases by 80% in female adult *Schistocerca gregaria* within 4 h after injection of a corpora cardiaca extract, and returns to its original level within 24 h. Conversely, fat body protein synthesis is reduced by 80% in female adult *S. gregaria 8* days after neurosecretory cell cautery or allatocardiacectomy. Allatectomy, alone, decreases fat body protein synthesis to a level intermediate between neurosecretory deficient animals and untreated animals. This latter result indicates that the neurosecretory system acts directly on fat body protein synthesis and not indirectly via an allatotropic effect on JH secretion. Total protein in the fat body, hemolymph and ovary of adult female *Melanoplus sanguinipes* is 50% lower after neurosecretory cell cautery than after allatectomy, and vitellogenesis is prevented. Ovarian proteins synthesized by the fat body are reduced in amount after either allatectomy or neurosecretory cell destruction but reappear after treatments with a JH analogue. In addition, nonovarian hemolymph proteins are absent after neurosecretory cell destruction. The fact that neurosecretory cell destruction reduces both ovarian and non-ovarian proteins suggests that the neuroendocrine system regulates two aspects of fat body protein synthesis. First, a neuroendocrine, allatotropic hormone stimulates JH secretion which affects the formation of specific ovarian proteins. Second, a neuroendocrine regulation of general protein synthetic capacity in the fat body lowers the producion of all fat body proteins.

A direct action on protein synthesis by neurohormones is indicated by the finding that fat body synthesis for five out of six hemolymph proteins, is increased by implants of brains or corpora cardiaca in decapitated *Leucophaea maderae* females. Of the proteins that respond to neurosecretions, four respond to brain implants and three respond to corpora cardiaca implants.

These results suggest that, in addition to neuroendocrine effects on general protein synthesis, specific neurohormones may also regulate the formation of specific proteins.

Most studies on the neurohormonal regulation of fat body protein synthesis have been general and not concerned with regulation of specific proteins. Scheurer's studies demonstrate that greater definition is needed in future investigations, and relationships should be considered between specific neurohormones and the formation of specific proteins. Neuroendocrine controls of fat body protein synthesis are generally unclear and have been largely ignored up to this point with major emphasis focusing on ecdysteroids and JH. However, neuroendocrine regulations will assume greater importance in the future as individual neurohormones are isolated and characterized and defined studies undertaken on specific effects.

Ecdysteroids exert both general and specific effects on fat body protein synthesis. Treatment of 7-8-day-old larvae of *Calliphora erythrocephala* with 20-OH-ecdysone causes a 35% increase in fat body protein synthesis after 4 h. All fat body proteins respond to the hormone and increase to the same degree with no specific effect being noted. This generalized increase in protein synthesis agrees with observations that ecdysteroids stimulate rRNA formation in the fat body and enhance the overall protein synthetic capacity of the tissue.

Few studies have related ecdysteroids to the synthesis of any specific fat body proteins. Recently, however, the synthesis of five specific proteins (P_{1-5}) was identified for the fat body in response to high ecdysteroid titers in third-instar larvae of *Drosophila melanogaster*. Of these proteins, P_1, P_3 and P_5 are newly translated from fat body mRNA in response to high ecdysteroid levels. Furthermore, cloned DNA probes show that 20-OH-ecdysone causes a 50-fold increase in the amount of poly(A)-RNA transcripts for P_1 in only the late third-instar larvae. P_2 is synthesized and secreted by the fat body when ecdysteroid titers are low, but P_2 is reabsorbed by the fat body and processed to form P_4 in late thirdinstar larvae when ecdysteroid titers increase. This indicates that P_2 formation occurs independent of ecdysteroids, but that P_2 processing to P_4 may be ecdysteroid-regulated. Lepesant, J.-A. *et al.* conclude that there are three effects by 20-OHecdysone: (1) gene specificity-induction of a single gene (P_1); (2) tissue specificity — only the fat body responds; and (3) stage specificity — the ecdysteroid effect on P_1 formation occurs only in the late thirdinstar larva and is not inducible in earlier larval instars.

Although the previous study demonstrates the formation of specific proteins by the fat body in response to ecdysteroids, it does not identify these proteins. Acid phosphatase is a specific protein that appears in response to 20-OH-ecdysone. Acid phosphatase occurs in adipocytes in association with the Golgi complex or in vacuoles, and is found in protein storage granules as pupariation nears in 4-day-old, third-instar larvae of C. *erythrocephala*. However, ligation blocks the formation of protein storage granules, and although treatment with 20-OH-ecdysone fails to stimulate the formation of granules in ligated larvae, acid phosphatase activity is nonetheless induced. Hence the appearance of acid phosphatase activity is a specific response to ecdysteroid and is related to, but independent of protein granule formation.

Larval-Specific Storage Proteins

The fat body synthesizes and secretes large amounts of a specific protein during immature life in several insect species. It has been speculated that these proteins serve as storage reservoirs for amino acids which are recovered and used later for the construction of new adult tissues during metamorphosis. The synthesis of these proteins appears endocrine-regulated. Although storage proteins have been identified in only a few insect species, they may be generalized and widespread.

The first larval-specific storage protein identified as a fat body product was calliphorin. Munn, *E. et al.* found that squashes of C. *erythrocephala* pupae contain large numbers of particles that yield a single electrophoretic protein band. This abundant protein was termed calliphorin, and the fat body was confirmed as its source. Calliphorin is not found in eggs and first appears in 2-day-old larvae. Calliphorin eventually constitutes 75% of the protein in the hemolymph of 6-day-old larvae, but it progressively disappears as the insect passes from the pupal stage into adult life. By day 3 of adult life. calliphorin constitutes only 10% of the total protein. Calliphorin is a single protein of molecular weight 5.28×10^5 daltons, composed of six subunits of 8.7×10^4 daltons each. Calliphorin is high in tyrosine, phenylalanine and methionine, but it contains very little cysteine or cystine. Some carbohydrate is present (0.4-0.5%). Calliphorin is synthesized only by fat body tissue from 3- to 5-day-old *C. erythrocephala* larvae; however, the production of calliphorin by adipocyte mRNA in a cell-free, wheat-germ system confirms that the calliphorin mRNA transcript is present in adipocytes between days 3 and 7 of larval age. This shows that calliphorin mRNA remains intact for 2 days after calliphorin synthesis has stopped. This is unusual since, normally, protein synthesis stops because of mRNA degradation. Although RNA remains high in larval adipocytes on day 7 the ribosomes disappear, and it is the loss of ribosomal structures, not RNA, that accounts for the cessation in calliphorin production.

Calliphorin accumulates rapidly in the hemolymph of *C. erythrocephala* larvae starting on day 3 and is reabsorbed by the fat body preparatory to pupariation when feeding ceases and wandering begins. Calliphorin is stored in protein storage bodies of adipocytes during pupariation and disappears rapidly during late adult metamorphosis. Presumably it becomes a source of amino acids for adult tissue formation and is undetectable in 2-day-old adult flies.

Although calliphorin is the best-known of the larval storage proteins, others have been identified in immatures of several other insect species. Two proteins that cross-react with anti-calliphorin antibodies are synthesized by mRNAs isolated from the fat body of larval *D. melanogaster*. These *in* vitro-translated proteins comigrate electrophoretically with two proteins that are extracted with anti-calliphorin from *D. melanogaster* larvae. The putative, larval-storage proteins from *D. melanogaster* larvae are termed drosopholin. Manducin is also a storage protein that comprises 80% of the hemolymph proteins in late-instar larvae of *Manduca sexta* and is composed of six subunits with a total molecular weight of 5.4 x 10^5 daltons. Manducin contains 20% tyrosine plus phenylalanine, 2% lipid and 3.5% carbohydrates. Finally, Kunkel, J. and Lawler, D. described a major protein found only in immature *Blattella germanica* cockroaches. A comparable larvalspecific protein is also found in immature mantids and termites. The larval protein of *B. germanica* varies 10-fold during the course of an instar, and reaches its maximum just before each ecdysis. Like calliphorin, the larval-specific protein disappears within 2 days in adults and, presumably, serves an amino acid storage function for histogenesis.

Endocrine regulation for the synthesis of stagespecific storage proteins is indicated but unclear. Sekeris, C. and Scheller, K. suggested that JH may regulate mRNA translation for calliphorin based on the timing for calliphorin production, but their suggestion is speculative and not supported by empirical evidence. The studies of Pan, *R. et al.* demonstrated inhibition of calliphorin synthesis by ecdysteroids. This inhibition agrees with the *in vivo* endocrine situation that shows a rise in ecdysteroid titers and a cessation in storage protein synthesis due to ribosomal destruction at the time of last instar wandering. The increase in ecdysteroid titer also stimulates adipocytes to take up the storage proteins from the hemolymph at the time of larval-pupal transformation in *Bombyx mori*. It can be speculated that since ecdysteroids suppress storage-protein synthesis, JH titers may exert antagonistic, stimulatory actions. However, in *B. mori,* JH suppresses storage protein synthesis, and the synthesis increases after allatectomy. Therefore the endocrine regulation of storage protein synthesis appears complex and must await further investigation for clarification.

Identification of endocrine effects on storage protein synthesis may provide an approach to elucidating differential cellular responses by adipocytes to hormones that regulate metamorphosis. The large quantities of the storage proteins that are produced suggest a large amount of fat body RNA is mRNA transcript for these unique proteins. Since the mRNA exists independently of storage-protein synthesis, regulation of synthesis must occur at the level of translation rather than transcription. This system, with its single, defined protein made in large quantity, can be a useful model for examining endocrine regulations of fat body function in immature insects just as vitellogenin synthesis provides a model for studying the mode of action for JH on the fat body of adult insects.

Protein Storage Granules

The formation of protein storage granules at various times during development is a feature of fat body biochemistry that appears universal for those insect species studied. In general, proteins are synthesized by the fat body and released into the hemolymph during the course of larval growth. At the onset of pupation some of the hemolymph proteins are sequestered back into the fat body and stored in large granules until metamorphosis, at which time they disappear. Presumably, like the larval-specific storage proteins, the proteins in the granules are a source of reusable amino acids for histogenesis during metamorphosis. The formation and disappearance of the granules appears endocrine-dependent with ecdysone having the greatest influence.

Protein granules form from several sources in the adipocytes of *Calipodes ethlius* larvae. Protein granules composed of either protein, or protein plus RNA, form by autophagy of cytoplasmic organelles just before the onset of pupation and as a first step in cytoplasmic reorganization of the adipocyte. A second type of protein granule forms by heterophagous uptake of hemolymph proteins into the adipocyte to form multivesicular bodies which fuse into granules. The appearance of injected plant peroxidase in the protein granules, and the decline in circulatory proteins at the time of heterophagous protein granule formation, both argue for the sequestration of hemolymph proteins into the granules. Fat body protein synthesis declines during the formation of heterophagous protein granules, and the granules contain little newly made protein.

Ecdysone secretion by the prothoracic glands corresponds to the formation of both heterophagous and autophagous protein granules. Heterophagous protein granule formation consists of a period of ecdysteroid-independent protein sequestration from the hemolymph into multivesicular bodies of the adipocytes, and an ecdysteroid-dependent coalescence of the multivesicular bodies to form protein storage granules. Likewise, fat body isolated from *C. ethlius* larvae prior to ecdysone secretion will not form autophagous granules *in vitro* unless 20-OHecdysone is present; however, adipocytes isolated from larvae after ecdysone secretion will initiate autophagy in a hormone-free medium.

Ecdysone injected into ligated abdomens from C. *ethlius* larvae stimulates pupation and the formation of protein granules after the absorption of specific hemolymph proteins. Three types of antigenic proteins are found in hemolymph from *C. ethlius* larvae and are synthesized and released by the larval fat body *in vitro*. Protein types I and 2 are always present in larval fat bodies, but type 3 appears in the adipocytes only during the last 24 h before pupation when protein granules are forming. Thoracic ligation before ecdysone secretion blocks pupation, the appearance of type 3 protein in adipocytes and the formation of protein granules. Protein types 1 and 2 are absorbed by adipocytes in ligated larvae, but type 3 is excluded from larval adipocytes and remains in the hemolymph. Therefore, it appears that type 3 protein is sequestered and stored in protein granules specifically in response to ecdysteroids. The nature of the type 3 protein is not defined, although it can be speculated that it may be a larval-specific storage protein.

Other insect species also exhibit protein storage granule formation in their adipocytes. Calliphorin synthesis and storage in fat body protein granules was discussed in the previous section. In *Hyalophora cecropia,* protein storage granules are composed of two proteins with molecular weights of 4.8 x 10^5 and 5.3 x 10^5 daltons. These proteins each have six subunits of 8.5 x 10^4 and 8.9 x 10^4 daltons, respectively. Both the molecular weights and their hexameric subunit structures are reminiscent of calliphorin and, like calliphorin, the constitutive protein is synthesized and secreted by the fat body of *H. cecropia* during early larval life and is reabsorbed and stored at the onset of pupation.

Acid phosphatase is associated with protein granules, which suggests that autophagous protein granules are derived from lysosomes. Locke, M. and Collins, J. described the structural transformation of multivesicular bodies containing acid phosphatase activity into protein granules in adipocytes of *C. ethlius*. In the fat body of *Calliphora erythrocephala* the protein granule fraction contains high acid phosphatase activity. Autophagic vacuoles with acid phosphatase activity and heterophagous vacuoles of absorbed hemolymph proteins fuse in the fat body of *C. erythrocephala,* confirming a partly lysosomal character and mixed origin for these protein granules. As discussed previously, the appearance of acid phosphatase activity in the fat body of *C. erythrocephala* depends on exposure to 20-OH-ecdysone.

Protein granules form in the fat body of *Drosophila melanogaster* larvae at the onset of pupariation by sequestration of hemolymph proteins in response to 20-OH-ecdysone. Transplantation of larval fat body into female flies also results in protein granule formation by the larval adipocytes. However, transplantation of larval fat body into male flies does not result in protein granule formation unless a larval ring gland (endocrine complex) is also implanted. Only the larval fat body forms protein granules in response to the ring gland; adult fat body is unresponsive. Furthermore, transplantation of ovaries into male flies along with larval fat body stimulates protein granule formation in the larval fat body. The ability of the ovaries to stimulate protein granule formation may be explained by the finding that ovaries of some Diptera secrete ecdysone, and the ovarian ecdysone stimulates the sequence of metamorphic events, including protein granule formation, in the transplanted larval adipocytes. This also explains the differences shown by larval fat body transplants in adult males and females.

Protein granules are also observed in adipocytes of adult, hemimetabolous insects. Generally, protein granules form in larvae immediately before pupation, and the proteins are presumably reused for histogenesis during metamorphosis. However, in adult insects the physiological role of protein granules is less certain. In adult male

Blaberus discoidalis cockroaches protein granules begin forming in adipocytes on the day of adult emergence with a maximum content present in day old adults. The granules disappear within 24-48h thereafter. These granules clearly show engulfment of mitochondria and endoplasmic reticulum, indicating an autophagous origin and function. The protein granules apparently engulf and destroy existing organelles and reuse their structural components for cytoplasmic reformation, which suggests a post-emergence, imaginal metamorphosis that reorganizes the cytoplasm of immature adipocytes into mature, imaginal cells. In female *Diploptera punctata,* protein granules form and disappear in inverse synchrony with reproductive cycles. Protein granules are abundant in adipocytes at adult emergence before vitellogenesis; disappear during vitellogenesis; then reform during oothecal gestation. In male *D. punctata* the initial granules disappear around day 4 of adult life, then reform and accumulate with no evident cyclicity. In the cases of both *B. discoidalis* and *D. punctata* the granules serve as reservoirs for proteins. In the first case the granules are apparently a protein source for reformation of imaginal cell structures in a manner comparable to their function in holometabolous insects. In the second case they act as a storage reservoir for proteins to be used for yolk formation.

Vitellogenic Proteins

A major function of the adult fat body in females is the synthesis and release of vitellogenic proteins for yolk formation during oocyte maturation. Vitellogenin synthesis accounts for 70-90% of the protein secreted by the mature fat body in female insects, and vitellogenin comprises nearly 90% of the yolk proteins. Furthermore, vitellogenin synthesis is endocrine-regulated, usually by JH. Particularly interesting is the fact that the fat body is not especially responsive to JH in immature insects, but it becomes a target tissue for the hormone in adults. The characteristics of endocrine-directed synthesis for large quantities of a uniform protein make the adult fat body useful as a model for studies on the cellular actions of JH. Vitellogenin synthesis has been reviewed several times recently.

Immunology and electrophoresis demonstrate that vitellin and vitellogenin are identical with a basic molecular weight of about 5.5 x 10^5 daltons. Vitellin contains 12.3% mannose, 1.3% glucosamine, and 9.6% lipids as neutral lipid, phosphatidylcholine and phosphatidylethanolamine. The vitellogenin molecule consists of two immunologically distinct components with molecular weights of 2.65 x

10^5 and 2.5 x 10^5 daltons. The synthesis of these two subunits is directed by two distinct genes. Both vitellin and vitellogenin of *Locusta migratoria* consist of eight separable polypeptides falling into two molecular weight ranges of either 1.4 x 10^5to1.05 x 10^5or6.5 x 10^4to5.2 x 10^4 daltons. These two categories of molecular weights represent the products formed by adipocytes processing the initial polypeptide products (2.5 x 10^5 daltons) from the vitellogenin genes. Variabilities in protein processing account for the two size ranges in the resulting eight polypeptide subunits that comprise the final vitellogenin molecule.

Imaginal adipocytes increase their cytoplasmic capacity for protein synthesis in response to the endocrine milieu that promotes vitellogenesis. Feeding often acts as a stimulus to the endocrine system to initiate vitellogenic activity. In flies and mosquitoes, protein feeding activates the endocrine systems and ribosomes, mitochondria and rough endoplasmic reticulum accumulate in adipocytes, produce protein-filled secretion granules, then regress. In *L. migratoria* the fat body stores nutrients for somatic growth during the first week of adult life. At about day 8, both rough endoplasmic reticulum and Golgi proliferate, and the tissue function changes from storage to protein synthesis and secretion. Not only are new proteins synthesized for vitellogenesis at this time, but proteins deposited in storage granules are mobilized by enzymes released from multivesicular bodies. Allatectomy prevents the conversion from the storage to the synthetic phase in the locusts, whereas corpora allata implants or treatment with JH analogs stimulate protein production. Protein granules are characteristic of immature adipocytes in newly molted female *Nauphoeta cinerea,* but they disappear 4 days after adult emergence when vitellogenin starts to appear. The proteosynthetic organelles become apparent about day 3, with peripheral adipocytes more active for vitellogenesis than centrally located cells.

Fat bodies in those insect species that have brief adult lives and do not feed may produce vitellogenic metabolites independent of hormonal regulations. A classic example of this is the saturniid silkmoths which lack adult mouthparts and contain fully formed eggs at adult emergence. The fat body of *Antheraea polyphemus* produces a female-specific protein during metamorphosis that is sequestered by the ovaries. JH does not stimulate synthesis or release of the female-specific protein, but increases the hemolymph titers of the protein by inhibiting ovarian maturation and preventing its uptake into the ovaries. In *Hyalophora cecropia,* vitellogenin accumulates in the hemolymph twice during metamorphosis. It first appears in the pharate pupa 6-10 months before oocyte development; it next appears in the second week of adult metamorphosis during tissue reconstruction. It is interesting to note that ovaries implanted into male silkmoth pupae also develop mature eggs. It is not clear whether this is due to uptake of general cytoplasmic proteins and metabolites discharged from pupal adipocytes during metamorphosis, or whether the ovaries supply a specific stimulus to the male fat body. Allatectomy of larval instars or pupae of H. *cecropia* fails to alter vitellogenin synthesis in the pharate adult, and injections of JH into isolated pupal abdomens do not stimulate vitellogenin formation. At the time of writing, no endocrine regulations of vitellogenic protein synthesis have been identified for saturniid silkmoths.

Vitellogenic protein synthesis is hormonally regulated in most insect species. Wigglesworth, V. first indicated a role for the corpora allata in egg maturation of *Rhodnius prolixus*. Following this initial report it became generally accepted that JH secreted by the adult corpora allata was the hormone regulating fat body production of vitellogenic proteins. Furthermore, JH was focused on in adult insects because it was thought to be the only major hormone present. Ecdysteroids were believed to be absent in adult insects because the prothoracic glands deteriorate shortly after the adult molt. However, because of recent discoveries concerning ecdysone secretion from ovaries and the action of ecdysone on fat body vitellogenin synthesis, the question of endocrine regulations of the fat body for ovarian maturation has been reopened amid controversy.

JH acts on the female fat body and ovaries to stimulate vitellogenesis. Vitellogenin appears in the hemolymph 40h after treatment of female adult *Leucophaea maderae* with JH-active analogs. This protein is not produced by fat bodies from males, nymphs of either sex or reproductively inactive females. Two JH-dependent vitellogenic proteins appear in the hemolymph of female adult *Periplaneta americana* at 4-5 days of adult age and accumulate in the hemolymph, after ovariectomy, to concentrations 20 times normal. Abdomens isolated from adult female *L. maderae* show an 18 h lag period before vitellogenin synthesis starts following injection of JH-1. The vitellogenin synthesis increases 50-fold by 72-96 h after JH treatment; then it starts to decline. JH is also necessary for ovarian uptake of the proteins, since injection of vitellogenin into allatectomized females does not, by itself, result in yolk formation.

JH appears to stimulate transcription. Fat body RNA doubles during the 8-day previtellogenic period in adult *Locusta migratoria* females and reaches its maximum on day 7. Vitellogenin synthesis starts an day 7 and reaches its peak on day 13 when vitellogenin comprises 60% of the total proteins synthesized by the adipocytes. JH treatment of allatectomized L. *migratoria* females normally produces maximum vitellogenin synthesis 72h later. However, a second application of JH within 10 days after the first renews vitellogenin synthesis after only a brief time lag due to the rapid appearance of polysomes that bear the vitellogenin transcript. This is explained by the finding that total RNA, especially rRNA, increases in response to the first exposure of JH, whereas only a small amount of RNA is produced in response to the second exposure to JH. The first exposure appears to stimulate transcription of vitellogenin genes along with genes for "bulk" RNA. The second exposure affects mainly vitellogenin gene transcription, while the majority of the ribosomal system is still intact and functional. Engelmann, F. demonstrated that actinomycin D blocks JH induction of vitellogenin synthesis, and existing vitellogenin mRNA transcripts persist for about 3 days. These findings suggest that the vitellogenin transcript is relatively short-lived, but ribosomes formed in response to the initial JH exposure remain available to resume protein synthesis in response to renewed JH induction of vitellogenin mRNA.

JH also increases endoplasmic reticulum formation in adipocytes. The rate of formation of endoplasmic reticulum increases by 3.5 to 5 times in JH-treated *L. maderae* females, and polysomes and heavy microsomes proliferate in reproductive females as compared to males or non-vitellogenic females. Endoplasmic reticulum formation increases the number of binding sites for polysomes during vitellogenin synthesis, and forms vesicles as part of the mechanism for vitellogenin secretion from the adipocyte.

During vitellogenin synthesis the newly synthesized proteins are released from the polysomes into the lumen of the endoplasmic reticulum which is then physically transported to the adipocyte surface for secretion.

In mosquitoes, adipocyte RNA rises measurably within 2 h after a blood meal and is increased by 3fold within 12 h. Vitellogenin synthesis is detected within 3 or 4 h after blood feeding, and peaks at 28 h.

Ovarian ecdysone regulates the onset of vitellogenin synthesis after blood feeding in *Aedes aegypti*. The ovaries secrete ecdysone as a prohormone which is converted by the fat body to 20-OH-ecdysone, the active hormone. Fat body vitellogenin synthesis is responsive only to 20-OH-ecdysone, not to ovarian ecdysone *per se*. Fat bodies from newly emerged females were unresponsive to 20-OH-ecdysone, and developed responsiveness only after exposure to JH.

The cellular action of 20-OH-ecdysone is unclear. Fong, W.-F. and Fuchs, M. found that a-amanitin and cordycepin, inhibitors of mRNA transcription and poly(A) mRNA formation respectively, do not prevent vitellogenin formation *in vivo*. However, actinomycin D, which inhibits transcription by blocking DNA, is inhibitory. In contrast, Kaczor, W. and Hagedorn, H. found that both aamanitin and cordycepin inhibit vitellogenin production by the fat body *in vitro*. The results of these two studies are contradictory and remain to be resolved. However, it should be pointed out that Kaczor, W. and Hagedorn, H. reported that their mosquitoes would not feed after α-amanitin treatment, and although Fuchs, M. and Fong, W.F. did not report trouble feeding their mosquitoes, they found that digestion was impaired. Hence the a-amanitin-treated mosquitoes studied by Fong, W.-F. and Fuchs, M. may have been abnormal if they were unable to digest their blood meal and receive the proper nutritional stimuli needed to trigger the sequence of in *vivo* endocrine events.

However, conflicting results have cast doubt on 20-OH-ecdysone as the principal regulatory hormone for vitellogenin synthesis in the fat body of mosquitoes. Physiological doses of 20-OHecdysone do not stimulate vitellogenin production in unfed *A. aegypti* females, and doses of hormone (0.05-5 μg) that exceed physiological levels (275 pg, fed females) by hundreds or thousands of times stimulate less vitellogenin formation than blood feeding or the presence of ovaries in fed females. In contrast, JH (12.5 ng) at nearly physiological levels stimulates 70-80% ovarian development in blood-fed, decapitated *A. aegypti* females. Finally, mature oocytes form in abdomens isolated from the autogenous mosquito, *Aedes atropalpus*, after treatments with 0.5 ng of JH-I followed by 4.8 ng of 20OH-ecdysone. Although only 50% of the oocytes develop the hormone doses are nearer to physiological levels. These results suggest that either hormone, alone, in high, non-physiological doses can stimulate the fat body to produce vitellogenic metabolites. However, the two hormones acting in concert in a properly timed sequence interact to increase the vitellogenic capacity beyond that of either hormone alone, and JH pretreatment sensitizes the fat body to 20-OH-ecdysone.

Finally, the ovaries may act as a "sink" to absorb vitellogenin and relieve a feedback inhibition of its further synthesis or release by the fat body. Engelmann, F. reported that the levels ofcirculating vitellogenin inhibit the release of further vitellogenin from the fat body of *Leucophaea maderae* but do not suppress its continued synthesis. However, Borovsky, D. found that ovariectomy of vitellogenic female *A. aegytpi* stops further vitellogenin synthesis after vitellogenin levels increase in the hemolymph by 4.5 times. Implants of vitellogenic ovaries reactivate the vitellogenin production, although similar ovarian implants fail to stimulate protein synthesis in previtellogenic, bloodfed females. These results suggest that there may not be an ovarian factor that stimulates vitellogenin formation. Rather, the ovaries may relieve an endproduct inhibition of vitellogenin synthesis orsecretion by removing circulating vitellogenin.

Although feedback inhibition may affect vitellogenin production by the mosquito fat body, the wealth of evidence supports endocrine regulations. The data suggest a timed sequence of cooperative interactions between JH and 20-OHecdysone for optimal vitellogenin synthesis. JH is secreted before blood feeding in young adult female mosquitoes, and appears to prime the adipocytes for vitellogenin synthesis when exposed, later, to ovarian ecdysone. JH priming of the fat body may involve activation of the C_{20}-hydroxylation enzymes for converting ovarian-derived ecdysone to hormonally active 20-OH-ecdysone, or it may involve the formation of cytoplasmic receptor proteins to increase adipocyte sensitivity to 20-OHecdysone. However, the cellular effects by the two hormones that result in enhanced vitellogenin synthesis remain obscure and should prove to be a fruitful area of future investigation.

In *Drosophila melanogaster,* both JH and 20-OHecdysone stimulate the synthesis of three yolk proteins. Treatment with a JH analog affects the synthesis and secretion of the vitellogenins by both the fat body and ovaries, whereas 20-OH-ecdysone affects only the fat body. This finding stresses the pre-eminence of 20-OH-ecdysone for stimulating fat body yolk protein synthesis in the Diptera.

Further support for the role of 20-OH-ecdysone in vitellogenin synthesis of Diptera is found with male *Sarcophaga bullata.* JH-I does not induce vitellogenin formation by fat bodies from male S. *bullata,* but vitellogenin is produced by fat bodies of 20-OHecdysone-treated males 3 days after emergence, if the flies are first fed a high-protein (liver) diet. Ovaries implanted into male flies fail to induce vitellogenin formation or to become vitellogenic even after injections of 20-OH-ecdysone. However, both of the latter effects may fail because of the absence of JH. JH may be necessary for the "primer" effect on the fat body to convert ovarianproduced ecdysone into 20-OH-ecdysone. Furthermore, lack of yolk formation may result from inability of the oocytes to absorb vitellogenin in the absence of JH.

The content of protein present in the fat body at adult emergence is a factor in autogeny. An autogenous strain of *Lucilia cuprina* was compared to wild-type individuals that were anautogenous. Protein stores in autogenous females are 20% larger at adult emergence and last longer than in anautogenous females. The increased protein stores are derived from the larval fat body which persists longer into adult life in autogenous flies than does the larval fat body of anautogenous flies.

An interesting and unique relationship exists in fleas concerning endocrine regulation of vitellogenin synthesis by the fat body. The fat body organization of the flea responds to the hormones that regulate reproductive cycles in the host rabbit. Exposure of the flea to estrogens, or luteinizing hormone, or feeding the flea on a pregnant or lactating doe or on newborn rabbits causes the fat body to organize and increase in size, and mating and reproduction ensue. Conversely, exposure to progesterones or feeding on nestling rabbits causes fat body deterioration and the suspension of reproduction. The fleas are sensitive to JH so that the vertebrate host hormones appear to interact with the endocrine system of the flea to regulate flea reproduction. In this way the fat body synchronizes flea reproduction with the availablity of host offspring for transference.

Diapause Proteins

Patterns of fat body protein synthesis change during insect dormancy. Fat body protein synthesis declines to only 3% of its active larval level during diapause in saturniid pupae. In female adults of *Leptinotarsa decemlineata* the fat body synthesizes in different ratios, at all times, both vitellogenic proteins and proteins related to diapause. Under long-day conditions two vitellogenic proteins are synthesized most actively by the fat body; with short-day photoperiods three diapause proteins predominate. Both types of proteins are present in either photoperiodic situation but the relative rate of synthesis for each type correlates to the photoperiod length. The vitellogenic proteins are secreted from the fat body, but the diapause proteins are stored.

JH influences the production of the vitellogenic proteins in *L. decemlineata* but not in the manner usual for most insects. JH stimulates by 32-fold the synthesis of vitellogenin-1 in fat bodies of dormant, short-day beetles with minimal synthetic activity. However, after diapause termination, vitellogenin-1 increases by 160 times and vitellogenin-2 by 120 times, even in beetles that have undergone allatocardiacectomy. Hence, although vitellogenin synthesis responds to JH when synthetic activity is depressed, JH is not needed to produce vitellogenin in diapauseterminating females. There is no vitellogenin synthesis response to 20-OH-ecdysone in diapausing females. Furthermore, synthesis of the diapause proteins is not affected by JH in diapausing beetles.

It might be noted that a JH dose of 50 pg per female was used to stimulate vitellogenin formation in diapausing *L. decemlineata*. This dose exceeds normal JH levels and could be forcing a non-physiological response which may be similar to the effects of high 20-OH-ecdysone doses on vitellogenin synthesis by the mosquito fat body. At present, the endocrine regulations for the synthesis of the vitellogenic and diapause proteins in *L. decemlineata* are unclear, and further research on this topic may identify new endocrine relationships with fat body protein synthesis.

JH Carrier Proteins and JH Esterases

The fat body is the source for a low molecular weight, JH-specific carrier protein in *Manduca sexta* and a high molecular weight lipoprotein that also binds JH. These proteins are synthesized in the fat body and secreted immediately. The low molecular weight carrier protein increases JH solubility in hemolymph and prevents degradation by non-specific esterases. The putative carrier protein of *Locusta migratoria* is a lipoprotein with a molecular weight of 220,000 daltons. In *M. sexta* the carrier protein is smaller with a molecular weight of 28,000-34,000 daltons, and less than 1 % of the high molecular weight lipoprotein is associated with JH. It may be that the high molecular weight JH-carrier protein reported for locusts corresponds with the non-specific lipoprotein of hornworms. In hornworms the large lipoprotein only acts as a JH-carrier protein in the unlikely event that the capacity of the smaller protein is exceeded by the quantity of available JH. There appears to be no apparent regulation of JHcarrier protein synthesis by the fat body and ample specific carrier protein exists in the hemolymph to bind more than the total amount of JH present at any given time.

In addition to producing a carrier protein that protects JH, the fat body is also the source for the JH-specific esterase that degrades circulating JH. JH esterase reaches a major peak of activity on days

3 to *4* of the last larval instar in *M. sexta* when the JH titer is declining in preparation for metamorphosis A second smaller peak of esterase activity occurs at the time of larval wandering.

The regulation of JH esterase by the fat body appears complex. Timed ligation studies suggest that JH esterase formation is neurohormonally controlled for the first peak of esterase activity in larvae of *Trichoplusia ni* and JH-controlled for the second peak of activity. JH treatments suppress JH esterase formation in larvae of *Galleria mellonella* but induce the activity in pupae of both *G. mellonella* and *T. ni*. However, further studies on *G. mellonella* suggest a brain neurohormone stimulates JH esterase production some time after 48 h of the last larval instar. Jones, G. *et al.* suggest that both stimulatory and inhibitory neurohormones regulate JH esterase activity.

The question of neurohormonal regulation of JH esterase is worthy of elucidation since it relates to regulations for metamorphosis and has a pragmatic potential for insect growth manipulation.

Protein Kinases and Cyclic Nucleotides

Protein kinases phosphorylate proteins. Frequently, the protein substrate is an inactive enzyme that is activated by phosphorylation and regulates processes involved in metabolism or cellular growth and differentiation. The protein kinases respond to endocrine stimuli via the system of cyclic nucleotide second messengers. In general, vertebrate protein kinases respond to CAMP. However, in the fat body of larval *Hyalophora cecropia* and *M. sexta* all protein kinases were reported to be activated by cGMP. cGMPdependent protein kinases appear to predominate in lepidopteran tissues, whereas cAMP-dependent protein kinases occur frequently in *Blaberus discoidalis* and *Drosophila melanogaster*. Protein kinases of locusts are sensitive mainly to cGMP but to some degree to CAMP, and they require a high Mgt^{+} level for optimal activity.

The relative levels of CAMP and cGMP may be significant to specific aspects of cell growth and differentiation, and it is interesting that the ratios of these two cyclic nucleotides change during metamorphosis. Both adenylate and guanylate cyclase are membrane-bound enzymes in adipocytes of *H. cecropia,* and they increase in activity at the larval-pupal transformation with little decrease in activity during pupal diapause. In *Bombyx mori,* adenylate cyclase activity is high at the start of the last larval instar; declines by *90%* during pupation; then increases by two to three times above its minimal, pupal level just before adult emergence. Guanylate cyclase changes inversely with adenylate cyclase in *B. mori* and parallels the ecdysteroid titers. Since most fat body protein kinases appear cGMP-dependent the data suggest that some aspects in the action of ecdysteroids on adipocytes may involve guanylate cyclase activation and second messenger effects. The reciprocity of adenylate and guanylate cyclase activities suggests that these enzymes may respond to hormones and control competing or antagonistic processes of development and metamorphosis by changing the relative ratios of the cyclic nucleotides.

Lipid Metabolism

Synthesis

A large number of articles report on the lipid composition of insects, but most of these are concerned with extracts of the whole body and are not specific for fat body. Since fat body is the major tissue depot for lipids, Kilby, B. states that studies on whole-body lipid composition also reflect fat body lipid composition without serious error. However, in keeping with the original focus of this review, only those studies dealing specifically with fat body lipids will be considered, with the emphasis on metabolism. The interested reader is referred to the comprehensive reviews of Gilbert, L. and Downer, R. for details on the composition and general metabolism of insect lipids.

Triglycerides constitute the main storage form for fat body lipids. Palmitate is esterified to triglycerides rapidly in the fat bodies of larval and adult saturniid silkmoths but to diglycerides in the pupal fat body. Triglyceride formation exceeds diglyceride formation by 3-fold in larvae, but diglyceride synthesis exceeds triglyceride synthesis by 2-fold in pharate pupae. The most common fat body fatty acids are palmitic, palmitoleic, stearic, oleic, linoleic and linolenic. Only saturated and monoenoic fatty acids are synthesized from acetate by the fat body. Insects cannot synthesize polyenoic fatty acids, but monoenoic fatty acids form by direct desaturation of saturated precursors. Microsomal enzymes show desaturation activity in the fat body of *L. migratoria* with stearate as the principal substrate and NAD as the preferred cofactor.

The fat body synthesizes phospholipid. Phosphatidylcholine is the main fat body phospholipid in fifth-instar *M. sexta* larvae and comprises 41% of the lipid phosphorus in this species followed by phosphatidylethanolamine at 30.6%. ATP and CTP both stimulate the formation of phosphatidylcholine in the fat body of larval *Phormia regina,* thus confirming the choline donor role for CTP in insect tissues. Phosphatidylethanolamine predominates in *P. regina* fat body and is formed by an exchange reaction with choline from phosphatidylcholine. The data indicate two kinase enzymes for the initial phosphorylations of ethanolamine and choline in the fat body of *P. regina* larvae. One kinase is cholinespecific; the other reacts with both choline and ethanolamine.

Lipogenesis from carbohydrate precursors occurs in the insect fat body. Radioactive label from [^{14}C]glucose appears in fatty acids as well as glycogen reserves in *H. cecropia*. The fat body of *Periplaneta americana* contains cytosolic ATP-citrate lyase for the generation of acetyl CoA from carbohydrate-derived citrate that is released by mitochondria into the cytosol. The resulting acetyl CoA is then converted by acetyl CoA carboxylase to malonyl CoA by the addition of CO_2 from bicarbonate. In *Locusta migratoria,* lipogenesis from carbohydrate stores increases around day 6 of adult life during imaginal somatic growth. Both trehalose and glucose are converted to fatty acids via citrate and ATP-citrate lyase which increases on day 4 of adult life.

Although lipogenesis from carbohydrates occurs readily, glyoxylate cycle enzymes are absent in insects and there is no evidence for conversion of lipids to carbohydrates. In *H. cecropia* pupae there is no indication of fatty acid conversion to either glycogen or trehalose in adipocytes.

Mobilization

The fat body forms diglyceride and releases it as the major circulatory lipid. Triglyceride content declines in the fat body of young adult *P. americana* due to hemolymph diglyceride formation from triglycerides composed mainly of saturated fatty acids. Fat bodies from *Hyalophora cecropia, P. americana* and *Melanoplus differentialis* prelabeled *in vivo* with [^{14}C]palmitate subsequently release the ^{14}C label in diglycerides when the isolated tissues are incubated in hemolymph. Diglyceride formation and release is energy-dependent and is blocked by inhibitors of energy production; free fatty acid release increases after energy inhibition. Fat body from chilled *H. cecropia* prelabeled with [^{14}C]glycerol and [^{3}H]palmitate release a higher ratio of ^{3}H/^{14}C than is stored in the fat body. The differences in fatty acid composition of released and stored glycerides suggest that fat body lipid release is a directed event for glycerides with a particular fatty acid composition. The glycerides for release are synthesized by an energydependent process specificially from a separate pool of stored lipids (triglycerides). Inhibition of energy production blocks specific glyceride formation but still permits the release of the hydrolyzed free fatty acids.

In general, triglycerides are the storage form for fat body lipids, while diglycerides are the mobile hemolymph lipids. Exceptions exist, however, that raise questions about the generality of this statement. Triglycerides are the primary form of lipid released by the fat body of *Galleria mellonella* larvae when incubated *in vitro* in dilute hemolymph. P. *americana* fat body releases 85% free fatty acids when incubated in a medium containing bovine serum albumin. Both free fatty acids and triglycerides are released *in vitro* if the medium contains a minimum of 20% hemolymph, and triglyceride release is optimal in a medium of 50-100% hemolymph. Chang, F. and Friedman, S. surveyed the form of lipid released from the fat body in several insect species at various growth stages. Although the principal hemolymph lipid of *Manduca sexta* is diglyceride, *in vitro* fat body tissue from pupae or pharate adults releases mainly free fatty acids, and larval fat bodies fail to release lipids into any medium. Adult fat body releases either free fatty acids *(Leucophaea maderae)* or diglycerides *(Schistocera gregaria)*, and pupal fat body from *H. cecropia* releases free fatty acids, diglycerides and triglycerides. These findings support the conclusion that diglycerides are the predominant form for lipids circulating in the hemolymph, but they suggest that lipids may be released from the fat body in several forms which are converted to diglycerides during circulation.

The fat body exhibits lipase activity for degradation of the storage triglycerides into circulatory lipids. Triglyceride hydrolysis in the fat body of P. *americana* produces diglycerides in the 1,2 configuration. In the southern armyworm, *Prodenia eridania*, the adult fat body contains lipases for mono-, di- and triglycerides. Presumably the glyceride stores in adipocytes of adult *P. eridania* are hydrolyzed to free fatty acids which are released and circulated to the flight muscles for use as a flight fuel. Two major forms of lipases occur in the fat body of *L. migratoria*. An acidic lipase (optimum pH 5-6) degrades mono-, di-, and triglycerides and is present in both microsomal and soluble forms, and a monoglyceride lipase is present in either a soluble form with an optimum pH between 7 and 8.5 or a microsomal form with activities up to pH 9.

Circulating diglycerides are protein-bound and the lipoproteins are products of the fat body. The major lipid carrier is a globulin-like protein that combines with cholesterol, phospholipids and carotenes in addition to diglyceride and has a molecular weight of 700,000 daltons. This carrier protein binds 44% of the lipid. A second minor protein (500,000 daltons) lacks the carotene moiety and binds only 10% of the lipid. Lipoproteins comparable to those found in the hemolymph are produced by the fat body *in vitro* from fatty acid and amino acid precursors, and lipid-free extracts of the fat body can incorporate [^{14}C]palmitate into existing fat body apo-lipoproteins. Puromycin blocks the *in vitro* incorporation of labeled amino acids into the lipoprotein, but it does not inhibit incorporation of the lipid moiety. This indicates *de novo* synthesis of the protein moiety by the fat body. Diglyceride conjugation to the apolipoprotein occurs within the adipocyte since diglyceride will not bind to free hemolymph proteins *in vitro*.

Ketone bodies are formed and released by the fat body as an energy source for the peripheral tissues in addition to glycerides and fatty acids. Both β-hydroxy-β-methyl glutaryl-CoA synthase and β-hydroxy-β-methyl glutaryl-CoA lyase have high activities in the fat body of *P. americana*. Acetoacetate is the major ketone in the hemolymph and the flight muscle, and hydroxybutyrate is the major fat body ketone. The ketone bodies are produced from the fat body and used as oxidative substrates by the fat body, flight muscles and testes. All three tissues use acetoacetate in preference to hydroxybutyrate. During starvation the levels of acetoacetate and hydroxybutyrate increase, not because of increased synthetic activity, but because of changes in the balance of the overall rates of production, use and excretion. Starvation reduces the oxidative capacity for ketone bodies in the flight muscles and fat body and results in their accumulation.

Endocrine Regulation

Lipids accumulate in the fat body of many insect species during early adult life but are released for vitellogenesis in response to JH. JH depresses adipocyte lipid synthesis but stimulates ovarian lipid synthesis. Allatectomy results in fat body hypertrophy because triglyceride turnover decreases. Furthermore, lipogenesis from carbohydrates may increase after allatectomy due to enhanced activities for ATP-citrate lyase and malic enzyme. These changes in adipocyte metabolism are direct responses to the absence of JH following allatectomy since fat body hypertrophy does not occur after ovariectomy. Yolk precursors are still mobilized after ovariectomy, but the released precursors accumulate in the hemolymph and may suppress further precursor formation. In addition, the increased osmotic pressure of the hemolymph due to the accumulated precursors may deter further feeding, especially in Diptera. Thus, although the precursors are not removed by the ovarian "sink", they do not accumulate in the fat body, and they suppress additional nutrient intake so that fat body hypertrophy does not result.

The release of lipid from the fat body for transfer to developing ovaries is accompanied by production of specific carrier proteins. Two proteins, designated A and B, are secreted by the fat body and bind to diglycerides in *Locusta migratoria*. The A-protein is a diglyceride-carrying lipoprotein present in the hemolymph throughout vitellogenesis. The B-protein is a female-specific lipoprotein that appears during vitellogenesis and is probably identical to vitellogenin. Protein A was not JH-regulated and did not remain in the oocyte after releasing its lipid.

Factors in the corpora cardiaca influence lipid oxidation and content in the fat body. The addition of corpora cardiaca extracts increases palmitate oxidation by 35% in the *in vitro* fat body from *L. maderae*. Carbohydrate and lipid stores accumulate in the fat body of adult *Pyrrhocoris apterus* following allatectomy but not after allatocardiacectomy. It is suggested that allatectomy disrupts reproduction-related metabolism in *P. apterus* and unused yolk precursors accumulate in the adipocytes. In contrast, the corpora cardiaca regulate "trophic" (digestive and basal) metabolism and allatocardiacectomy reduces digestion and the conversion of nutrients into metabolite stores.

An adipokinetic hormone (ADKH) from the corpora cardiaca stimulates the fat body to release diglycerides and increases the oxidation of lipids in preference to carbohydrates by the flight muscles in adult *Schistocerca gregaria* locusts. ADKH is a decapeptide and a product of the glandular lobe of the corpora cardiaca. Corpora cardiaca from *Tenebrio molitor* and *S. gregaria* are interchangeable for stimulating lipid mobilization in these two species. However, injection of *Periplaneta americana* gland extracts into *Locusta migratoria* causes lipid release, but fat body of *P. americana* does not release lipids in response to either its own corpora cardiaca or to glands from locusts. Instead, cockroach glands act on cockroach fat body in the reverse of the locust glands and cause lipid uptake and storage.

It is particularly interesting to note that, although ADKH is one of two chemically defined insect neurohormones and much of its physiological significance is known, the means by which ADKH mobilizes fat body lipids is not clearly established. There is no measurable activation of fat body lipase in response to ADKH although

Cat and hemolymph proteins are needed for the lipid mobilization effect and cAMP appears involved. Fat body cAMP content increases with 5 min after corporate cardiaca extract administration and gives a linear dosedependent response for glandular doses up to 0.01 gland pair. ADKH administration also results in an increase in fat body cAMP-activated protein kinase activity *in vitro* within 3.5 min. Incubation of fat body with cAMP *in vitro* activates lipase activity and increases diglyceride production by 2 - fold. This sequence of*in vitro* events argues for the activation of fat body lipases in response to ADKH activation of the cAMP-second messenger system. However, it was not confirmed experimentally that ADKH did, in fact, stimulate lipase activity. Furthermore, only adult *L. migratoria* between 3 and 35 days old give a lipid mobilization response to ADKH even though fat body cAMP increases after ADKH treatment for all ages tested, and lipid is always available.

The inability of locusts to respond at all ages to ADKH may result from the absence of a particular form of diglyceride-carrier protein. The usual diglyceride-carrier protein (A) is converted to a higher molecular weight protein (A+) after injections of ADKH-active corpora cardiaca extracts (Mwangi, R. and Goldsworthy, G., 1977b). The A protein, by itself, is unable to load diglyceride normally. As A+ forms, a lower molecular weight protein (C) disappears. Incubation of A+ in the presence of fat body gives rise to both A and C. Presumably, A combines with C in response to ADKH to form A + for lipid transport. Young (< 3 days old) locust adults that are not responsive to ADKH, lack protein C and cannot form A+ after treatment with ADKH.

In conclusion, the ADKH response appears to consist of two components. Somehow ADKH stimulates the cAMP-second messenger system in adipocytes to cause lipid release and, although how this occurs is not demonstrated, lipase activation appears involved. At the same time two proteins, A and C, are needed in the hemolymph to form the carrier protein for the mobilized diglycerides. Presence of the C protein may not be ADKHrelated but is a normal, timed, developmental event. Possibly the timed formation of these specific carrier proteins is regulated by other, presently unknown endocrine stimuli related to adult maturation.

Carbohydrate Metabolism

General Metabolism

Like lipids, carbohydrates are important circulatory metabolites synthesized and released from the fat body for use as an energy source by peripheral tissues. Friedman, S.; Chippendale, G.; and Steele, J. have reviewed aspects of the role of the fat body in carbohydrate metabolism.

As found in other animals, the normal glycolytic pathways *occur* in insect adipocytes. Properties and activities are described for several major glycolytic enzymes from the fat body: hexokinase, phosphofructokinase, aldolase and pyruvate kinase. Like the flight muscle, a glycerophosphate shuttle is active for the oxidation of cytoplasmic $NADH_2$ and lactate dehydrogenase activity is minimal. The TCA cycle and electron transport enzymes are described for adipocytes of S. *gregaria*. In fact, there is little reason to doubt the presence in the fat body of all the common enzymes for glycolysis and the TCA cycle.

The pentose phosphate shunt may be a significant route for glucose degradation in the fat body. Phosphofructokinase has the least activity of any glycolytic enzyme in adipocytes of *P. americana* and high activities by glucose-6-phosphate dehydrogenase and aldolase suggest the pentose phosphate pathway is a significant alternative for triose formation in adipocytes. Generally, however, the pentose phosphate pathway is considered minor for carbohydrate degradation in insects with less than 30% of the glucose being metabolized by this route. The pentose phosphate pathway is used mainly to produce $NADPH_2$ for fatty acid synthesis or for insecticide detoxification by the mixedfunction oxidase pathway after sublethal poisoning.

In addition to the usual routes for carbohydrate degradation, the fat body contains enzymes for gluconeogenesis. Pyruvate carboxylase (mitochondrial), content increases within 5 min after corpora carphosphoenolpyruvate carboxykinase (cytosolic), fructose- 1,6-diphosphatase (cytosolic) and glucose6-phosphatase (microsomal) are all key enzymes in the fat body of *P. americana* for the resynthesis of glucose from pyruvate.

Glycogen is the principal storage form for carbohydrate in the insect fat body and may comprise as much as 10-25% of the dry weight of the tissue. Glycogen is synthesized by glycogen synthase which is a bound enzyme stimulated by glucose-6phosphate and Mgt+.

Phosphorylase degrades fat body glycogen, and its activity is affected by several regulatory mechanisms. Inactive phosphorylase *b* is converted to active phosphorylase *a* by a kinase which is, itself, activated by ATP, Mg^{2+}, and Ca^{2+} but not by Pi. As in the vertebrate liver, AMP increases the activity of fat body phosphorylase *b*; however, the increase in activity of fat body phosphorylase *b* is not strictly AMP-dependent. Locust phosphorylase does not require AMP for activity, and all samples tested gave the same 75% increase in activity over a 200-fold range of AMP between 0.005 and 1 mM.

Furthermore, trauma to the fat body activates phosphorylase. Standing *in vitro* increases the active enzyme level from 1 % to 30% in tissue from diapausing silkmoth pupae. Since AMP does not increase in the isolated tissue, the increase in phosphorylase activity is probably enzymically controlled by phosphorylase kinase. In support of kinase activation, both cAMP and phosphorylase *a* increase in the tissue after excision. Hence, trauma activation of phosphorylase may result from increased cAMPdependent, phosphorylase kinase activity. Traumainduced activation of phosphorylase results in trehalose synthesis.

Exposure to cold also activates phosphorylase but with the formation of glycerol rather than trehalose. Pupae held at 4° show an increase in fat body phosphorylase from a basal level of 8-10% active form up to 40% active form within 1-2 h. This increased activity is maintained for 30 days along with glycerol accumulation in the hemolymph. The rate of glycerol production from glycogen by the fat body at 6° increases by 2-2.5 times above its rate at 25° and decreases when shifted back to 25°.

Cold activation of fat body phosphorylase occurs by a different mode of action than does trauma activation. Fat body held *in vitro* at 25° shows the trauma activation of phosphorylase (40%) that reverts to lower activity (7-10%) within an hour, whereas phosphorylase activity persists at high levels (80%) in fat bodies held *in vitro* at 4°. Trauma activation correlates to increased levels of fat body cAMP and phosphorylase *a*; no increases are noted for cAMP or cGMP during cold activation of phosphorylase. Therefore, how cold activates fat body phosphorylase remains unresolved, but it does not appear cAMP-related.

Phosphorylated glucose intermediates regulate the synthesis and degradation of glycogen in insect adipocytes. Glucose-6-phosphate stimulates glycogen synthase in *Hyalophora cecropia* fat body and inhibits phosphorylase in the locust fat body. These latter reports suggest that glucose-6-phosphate levels determine whether glucose is converted to either stored or mobile carbohydrate. If ample dietary glucose is available and trehalose levels are high, then accumulating levels of glucose-6-phosphate suppress further phosphorylase degradation of glycogen stores and stimulate glycogen synthase to store the accumulating glucose. Conversely, as supplies of glucose-6-phosphate are depleted by trehalose synthesis or glycolysis, phosphorylase is activated and new phosphorylated glucose intermediates are produced by glycogen degradation for either additional trehalose formation or energy production.

Trehalose Syntheses and its Regulation

Trehalose is the major circulatory carbohydrate of most insects and is synthesized solely by the fat body. Trehalose was first identified as the principal hemolymph sugar in 10 species of insects representing five orders. In Lepidoptera, trehalose generally ranges from 200 to 1500 mg% in the hemolymph and constitutes 90% of the circulatory carbohydrates. The trehalose biosynthetic pathway was determined in the locust fat body and consists of the conjugation of glucose6-phosphate and uridine diphosphoglucoside (UDPG) to form trehalose-6-phosphate which is subsequently dephosphorylated to trehalose.

The ability of the fat body to produce trehalose may differ between growth stages in some species. For example, glucose is the primary carbohydrate in the hemolymph of larval *Phormia regina* even though trehalose can be detected within larval adipocytes. Trehalose is not detected in the hemolymph until pupariation.

Hemolymph trehalose levels are regulated by activation and feedback inhibition of key enzymes in the trehalose synthetic pathway. In *H. cecropia,* trehalose inhibits fat body trehalose-6-phosphate synthase by allosteric feedback inhibition to prevent the build-up of excessive trehalose levels. Trehalose-6phosphate synthase is the key enzyme for trehalose synthesis and is activated by Mgt+ which lowers its K_m for the glucose-6-phosphate substrate. Trehalose competes with glucose-6-phosphate for an allosteric site on trehalose-6-phosphate synthase, and as trehalose concentrations approach 50 mM, trehalose strongly inhibits its synthetic enzyme. Just as trehalose inhibits its formation by inactivation of the key synthetic enzyme, so too, trehalose stimulates a phosphohydrolase in *P. regina* fat body to release Pi from glucose-6-phosphate. Hence, increasing concentrations of trehalose stimulate the degradation of one of the main precursors for further trehalose biosynthesis.

Trehalose synthesis is also regulated by the relative affinities of trehalose-6-phosphate synthase and glycogen synthase for their common substrate: UDPG. In *H. cecropia* fat body, the K_m values for UDPG are 0.3 mM for trehalose-6-phosphate synthase and 1.6 mM for glycogen synthase. The lower K_m of trehalose6-phosphate synthase for UDPG indicates that, given the competition for conversion of UDPG to either trehalose or glycogen, the trehalose pathway is most active with glycogen being formed only after trehalose levels have reached the concentrations needed for feedback inhibition of trehalose-6phosphate synthase. The relative importance of the two competing pathways is confirmed *in vivo* in *Periplaneta americana* by injection of glucose. Injected glucose is converted to trehalose during the first 10-20 min; after 20 min, glycogen synthesis predominates. During the first 20 min circulating trehalose levels increase, but thereafter they remain constant. These results demonstrate that the production of circulating carbohydrate takes precedence over carbohydrate storage. Only after the needs for circulating carbohydrate are satisfied does excess carbohydrate convert to glycogen stores.

An additional regulation of trehalose synthesis occurs by differential permeability of the adipocyte plasma membrane. In *H. cecropia* larvae, trehalose penetrates readily into adipocytes; however, in pupae, trehalose is restricted from passing through the adipocyte plasma membrane. This impermeability to trehalose is reversed in early pharate adults. Trehalose that is synthesized is retained in the pupal adipocytes and suppresses further trehalose synthesis by feedback inhibition. Therefore, low hemolymph trehalose levels are maintained in *H. cecropia* pupae. Furthermore, Mg^{2+} reverses the feedback inhibition by trehalose on trehalose-6-phosphate synthase. The loss of Mg" from adipocytes during the larval-pupal transition enhances the feedback inhibition, and blood trehalose levels correlate with Mg^{2+} levels in the adipocytes during pupal life.

Endocrine Regulation

Hemolymph trehalose levels are regulated by a corpora cardiaca neurohormone designated, recently, as trehalagon. Injections of corpora cardiaca extracts elevate hemolymph trehalose at the expense of fat body glycogen in *Blaberus discoidalis* and *P. americana*. The gland extracts increase fat body phosphorylase activity by activation of adenyl cyclase to form cAMP. Adenyl cyclase activity increases by 3-fold and doubles the cAMP content in the fat body within 5 min. These data suggest that trehalagon activates fat body phosphorylase via the cAMP second messenger system to degrade glycogen and produce the intermediates for trehalose synthesis.

Not all insects have a neuroendocrine factor that stimulates the production of trehalose by the fat body. For example, there is no evidence for a neuroendocrine factor affecting fat body phosphorylase in *H. cecropia*, nor does fat body phosphorylase of *Hyalophora cecropia* respond to extracts of glands from *P. americana* that stimulate *P. americana* phosphorylase. However, in *Manduca sexta,* another lepidopteran, corpora cardiaca extracts elevate hemolymph trehalose in the usual dosedependent manner.

In cockroaches, corpora cardiaca extracts activate fat body phosphorylase at any time, and a maximum of activity is reached within 10 min after injection. Injection of 0.025 corpora cardiaca increases hemolymph trehalose by 236% within 30 min in P. *americana* and maintains it at this level for 5 h, while the fat body glycogen content declines from 5.4 mg to 0.9 mg within I h.

In Diptera, responsiveness to corpora cardiaca extracts by fat body phosphorylase is a function of the nutritional state of the fly. Corpora cardiaca extracts increase hemolymph trehalose only in *Phormia regina* that are starved for 24 h. This suggests that trehalose levels are normally maintained in flies by synthesis from nutrient-derived glucose. Only after starvation depletes the nutrient glucose does the fly rely upon trehalagon to insure the maintenance of hemolymph trehalose levels by the degradation of fat body glycogen reserves.

Finally, adult *Locusta migratoria* fail to show a clear trehalagon activation of phosphorylase during the first 6 days of adult life because glycogen stores are minimal. After 6 days of adult life, when somatic growth is finished, corpora cardiaca extracts elevate hemolymph trehalose levels because glycogen stores have accumulated.

These results demonstrate that phosphorylase activation by trehalagon is a function of the peculiar biology and physiology of a particular insect species or order. Insects that feed frequently, such as flies, have an ample supply of diet-derived glucose for use directly as trehalose precursors. Therefore, they require trehalagon only intermittantly as a back-up system to mobilize fat body glycogen stores in cases of starvation, or if the digestive source is too slow to meet the demands of rapid trehalose consumption. In contrast, the cockroach constantly uses the endocrine system to maintain trehalose levels by glycogen degradation. Finally, in locusts, sustained flight uses lipids for energy. Therefore, as discussed previously, carbohydrates are converted to lipids and lipid stores accumulate, hence glycogen stores are small and phosphorylase activity and its regulation are generally insignificant.

In most cases trehalose levels are high in insect hemolymph in comparison to glucose levels in mammalian blood. This is probably an adaptation by insects to the open circulatory system which distributes hemolymph metabolites to sites of consumption, in part by diffusion along concentration gradients. This is inefficient relative to the closed circulatory system and directed blood flow of vertebrates. Therefore a constant, high level of circulating carbohydrates is required in insects to compensate for circulatory inefficiency and to fulfill immediate emergency energy demands in peripheral tissues (especially in leg and flight muscles). Since trehalose is an inert disaccharide it provides two molar equivalents of glucose for use in metabolism relative to one molar equivalent of osmotic effect. The normally high concentrations of trehalose are sufficient to meet immediate needs. However, when depleted by consumption or starvation, trehalagon insures a rapid recovery by trehalose to its naturally high levels through the degradation of fat body glycogen reserves.

ADKH also activates fat body phosphorylase, but it does not appear to account for all of the hypertrehalosemic activity in corpora cardiaca extracts. Purified locust ADKH increases hemolymph trehalose in cockroaches, but only at doses 10 times greater than those needed for hyperlipemia in locusts, Corpora cardiaca extracts activate fat body phosphorylase in the locust, but treatments with synthetic ADKH indicate that all of the glandular activity is not accounted for by ADKH. Furthermore, extracts of corpora cardiaca storage lobes also activate locust phosphorylase but have no lipemic effect. Likewise, corpora cardiaca from M. *sexta,* which lack an ADKH effect also stimulate phosphorylase activity in the locust fat body. In locusts the hypertrehalosemic effect and the adipokinetic effect remain connected throughout purification. However, in *Periplaneta americana* both physiological responses occur but are associated with different chemical factors neither of which is identical to ADKH. These findings suggest that ADKH may stimulate phosphorylase to degrade glycogen, but a distinct hypertrehalosemic hormone exists in addition, in some insect species, that maintains hemolymph carbohydrates by regulation of fat body phosphorylase.

An excitation-induced hypertrehalosemic response occurs in insects. Handling or other types of excitatory stresses cause elevation of hemolymph trehalose even in head-ligated *P. americana,* due to effects on the fat body by octopamine, a biogenic amine. Hemolymph trehalose levels double within 15 min after injection of octopamine. Octopamine (10^{-6} M) elevates fat body cAMP within 10 min *in vitro*. Presumably, octopamine activates phosphorylase via cAMP for degradation of glycogen and formation of trehalose in the same manner as trehalagon.

A unique situation exists regarding amino acid metabolism in *L. decemlineata*. In this beetle the corpora cardiaca increase hemolymph glucose and decrease hemolymph alanine. The alanine is used by the fat body along with fatty acid-derived acetyl CoA to form proline. This is done by transaminating alanine with ketoglutarate to form pyruvate and glutamate. The glutamate is converted to proline and released into the hemolymph for transportation to the flight muscles where it serves as a flight fuel. The pyruvate derived by deamination of alanine is converted by malic enzyme to malate in the presence of CO_2 and $NADPH_Z$, and ultimately to citrate using fatty acid-derived acetyl CoA. The citrate is further converted to ketoglutarate and transaminated to glutamate for conversion to proline.

The proline-synthesizing ability of the fat body increases by 6-fold during the first 8 days of adult life in *Leptinotarsa decemlineata*. This corresponds to the time for development of flight capacity in the flight muscles. The ability to synthesize proline is sensitive to a corpora cardiaca extract which concomitantly decreases the hemolymph alanine. Synthetic locust ADKH also exhibits this effect. Furthermore, synthetic ADKH and extracts of both beetle and locust corpora cardiaca stimulate glucose release from the beetle fat body, whereas cockroach gland extracts do not. These results suggest that, as in the locust, ADKH controls fat body mobilization of metabolites that are used for flight in the Colorado potato beetle.

Uric acid

Urocytes are cells, associated with the fat body, that store uric acid - the principal end product of nitrogen metabolism in many insects. Presumably uric acid is stored in the urocytes as a form of excretion. Urocytes appear to be degenerate cells lacking distinct cytoplasmic organelles and are located centrally in fat body lobes relative to the more peripheral adipocytes. In cockroach tissue the urocytes are associated with mycetocytes which may assist in remobilizing the stored nitrogen (see section on symbiotes).

Uric acid formation and degradation occur readily and with great variability in the fat body. Uric acid stores increase in the fat body after starvation, but the addition of xanthine, the immediate precursor for uric acid, to the diet of fed P. *americana* does not stimulate urate accumulation. In both *Bombyx mori* and *Aldrichna grahami* (blowfly), increased dietary protein stimulates both fat body xanthine dehydrogenase activity and urate formation. However, dietary xanthine does not increase xanthine dehydrogenase activity in A. *grahami;* rather, additions of amino acids related to purine formation, or additions of adenine or hypoxanthine, increase xanthine dehydrogenase activity by up to 3 times. In *L. maderae,* [^{14}C]adenine is converted to [^{14}C]urate and stored in the fat body where it accumulates progressively in a protein-bound form with little turnover. Only 2% of the initial ^{14}C label from [^{14}C]adenine is excreted in the feces after 12 days. This indicates that the urate stores are quite stable.

Metamorphosis affects the capacity for uric acid synthesis and storage in the fat body. Lafont, R. and Pennetier, J. find that uric acid stores in the pupae of *Pieris brassicae* butterflies are converted to allantoic acid just before imaginal eclosion, and it is allantoic acid that constitutes the major endproduct for purine metabolism. At the transition from feeding to wandering in *Manduca sexta* larvae, there is also a transition from excretion of uric acid by the Malpighian tubules to storage in the fat body. The fat body stores uric acid during the first half of pupation in *B. mori,* but transfers it to the rectum for excretion in the meconium midway through pupation. In *Hyalophora cecropia,* urate storage in the fat body at the larval-pupal transition is accompanied by storage of K^+ derived from the integument. The fat body uric acid probably results from uptake of potassium urate, although uric acid stores accumulate separately from potassium.

In general, urate storage may regulate osmolarity by sequestration of Na' and K^+ as stored urate salts in *P. americana*. This lowers the need for the excretion of these cations along with concomitant water loss. By alternately sequestering or releasing Na' or K^+ from the fat body urate stores, hemolymph osmolarity can be regulated and stabilized.

Uric acid stores appear responsive to the endocrine status of the insect. Bodenstein, D. reported that uric acid disappears from the fat body of *Periplaneta americana* after allatocardiacectomy. Corpora cardiaca implants reverse this effect, but corpora allata implants are without effect. These findings suggest neuroendocrine regulations for either protein-amino acid metabolism or uric acid metabolism. In *M. sexta,* uric acid deposits increase in the fat body during the last larval instar, but the capacity for urate synthesis decreases by 85% between the late feeding larva and the newly formed pupa. This decrease in urate synthetic ability can be prevented by thoracic ligation of the larva before the increase in ecdysteroid titer associated with the larval-pupal ecdysis. The decrease in urate synthesis occurs in ligated larval abdomens after injection of 20-OH-ecdysone. The presence of the hormone causes urate synthesis to decline as it would normally at the time for pupation. This effect may be a result of adipocyte metamorphosis rather than a direct effect by the hormone on the enzymes for uric acid synthesis.

Intracellular symbiotes and metabolism

Associated with the physiological role of urate stores are the symbiotes found in fat body mycetocytes and their role in metabolism. The beststudied fat body symbiotes are bacteria-like organisms of cockroaches referred to generally as bacteroids. Bacteroids in the fat bodies of all cockroach species are of one type identified as *Blattobacterium cuenoti,* and they possess features of both the Grampositive and Gram-negative bacteria and of Rickettsia. These organisms are situated in the mycetocytes that are located centrally in the fat body lobes and are surrounded by urocytes.

The spatial arrangement of the mycetocytes and urocytes suggests functional interactions which are confirmed by studies with aposymbiotic insects. Aposymbiotic nymphs and adult males of *Blattella germanica* contain significantly more fat body uric acid than do normal animals. Newly emerged aposymbiotic adult females contain more fat body uric acid than normal females, but this situation reverses as normal adult females accumulate uric acid faster than aposymbiotic females. These results confirm that bacteroids affect uric acid stores, and indicate a causal relationship for the cellular arrangements within the fat body lobes.

Fat body symbiotes affect the growth and reproduction of their host. Female *B. germanica* exposed to high temperatures or fed antibiotics are incapable of transmitting their fat body symbiotes transovarially and produce aposymbiotic nymphs Aposymbiotic nymphs are unable to grow unless brewer's yeast is added to their diet; then they grow slowly and require 2 to 3 times longer to reach maturity. Adult aposymbiotic insects reproduce poorly in both sexes, and females produce abortive oothecae. Implants of normal fat body grow in aposymbiotic nymphs and improve the growth of the host insect; however, treatments of aposymbiotic nymphs with suspensions of free bacteroids show that the bacteroids do not invade mycetocytes or promote nymphal growth. Aposymbiosis also hinders reproduction in *Blaberus craniifer*. Two out of three protein bands diminish in the hemolymph of aposymbiotic females. One of the bands is the female-specific protein, and although synthesis of the missing proteins recovers after 2 weeks, resulting oothecae are distorted, sterile and expelled early. Nymphs that emerge often die. Therefore, depletion of the bacteroids somehow reduces fat body protein synthesis contributing to reproduction.

Bacteroid symbiotes affect uric acid metabolism in nymphs. Adipocytes have low guanase activity for the conversion of guanine to xanthine for uric acid synthesis in nymphal *Leuçophaea maderae*. In addition, nymphal adipoctyes have low urate oxidase activity so that uric acid cannot be degraded to other end-products of nitrogen metabolism. In contrast to the nymphs, both guanase and urate oxidase activities are high in adipocytes of adult *L. maderae* and in bacteroid symbiotes in the mycetocytes. Nymphs may rely on the symbiote enzymes for urate metabolism since heavy deposits of urates are found in the fat body of aposymbiotic nymphs. Symbiotes apparently also assist their host insect by recovering the nitrogen stored in the urates. Cultures of isolated bacteroids are reported to degrade uric acid through allantoin, allantoic acid, urea, glyoxylic acid and finally to pyruvate for entry into the TCA cycle. Thus, the urate nitrogen may be recovered back into the general metabolism. However, Brooks, M. questions the successful isolation of bacteroids and casts doubt on the results from studies reporting their use.

Uric acid is generally accepted as the main excretory product of metabolism in terrestrial insects. However, some doubt has been thrown on this doctrine by the finding that it is ammonia that is the most prevalent nitrogenous component in the excreta of *P. americana,* with no uric acid being present. Consequently, the finding by Pierre, L. that low urate oxidase activity in adipocytes of nymphal *L. maderae* correlates to large fat body urate deposits lends credence to the concept that uric acid is not always a nitrogenous waste product. Rather, uric acid may be a nitrogen storage form that can be recovered by the symbiotes for reuse in growth processes of the host insect. In support of this, Mullins, D. and Cochran, D. demonstrate that uric acid is consumed more rapidly as a nitrogen source when *P. americana* is fed a diet with a negative nitrogen balance. Females fed a negative nitrogen diet mobilize urate stores faster than males, and a portion of the urate nitrogen is used for egg formation and is recovered in the ootheca. Since ammonia is the major excretory product of this insect, the urate nitrogen can be recovered and used in protein synthesis for growth and reproduction.

Alternative roles for bacteroids are unclear, although it is believed that they may contribute to the general metabolism of the fat body. Several *in vitro* studies claim to have elucidated the metabolic action of bacteroids; however, Brooks, M. criticizes the techniques used to obtain the bacteroid cultures, and indicates that it is highly unlikely that the cultured organisms were derived from bacteroids. Therefore, studies with aposymbiotic insects may be useful for identifying biochemical capacities of bacteroids and their metabolic role *in vivo*.

Comparisons of normal and aposymbiotic insects suggest that isocitrate dehydrogenase, malate dehydrogenase, urate oxidase and guanine aminohydrolase are enzymes associated with symbiotes. Álso, succinoxidase and cytochrome c oxidase activities are greater in fat body of *Nauphoeta cinerea* when bacteroids are present as compared to aposymbiotic tissue. This suggests that bacteroids have mitochondrial properties which may contribute to the total energy levels in the tissue. Some vitamins may be supplied by bacteroids since normal fat body homogenates synthesize ascorbic acid but not homogenates of aposymbiotic tissue. Finally, evidence from aposymbiotic *Periplaneta americana* indicates that the micro-organisms are instrumental in ring cleavage of phenylalanine. In contrast to the previous results, malic enzyme appears to be an adipocytic enzyme unrelated to the presence or absence of bacteroids.

Although these studies suggest that bacteroids have a significant metabolic role, several cautions should be noted. It is quite likely that the bacteroids exhibit a variety of enzymic activities; however, in view of the broad metabolic functions of the adipocytes it is also likely that many of the enzymes ascribed primarily to symbiotes may also reside in adipocytes at levels greater than those determined. For example, if Brookes, M. is correct, then the micro-organisms in culture studies were not *Blattobacterium cuenoti,* and the results of culture studies are erroneous. This means that the only currently valid basis for assessing bacteroidal enzymes is by comparing enzyme activities in normal and aposymbiotic fat bodies. In this case it seems reasonable to question the side-effects on the adipocytic enzymes of the harsh procedures used to render the fat body aposymbiotic. These procedures may damage adipocyte enzymes and processes as well as destroying the bacteroids, thus making the affected process appear bacteroid-related.

CONCLUSIONS

The purpose of this review is to provide to the interested reader an indication of the breadth of the fat body's role in the normal biochemistry of the insect. Some areas of fat body biochemistry, for example detoxification mechanisms, were omitted intentionally since they will surely be covered in more depth by other speciality chapters in this series. Obviously, a focus on normal biochemistry was sufficient.

Although many studies on insect biochemistry involve the fat body, it is only recently that insect research has begun to take advantage of the unique features of the fat body and to design experiments using the fat body as a model based on these features. As an example are the studies on endocrine regulation of vitellogenin synthesis. The regulation of vitellogenin synthesis by JH promises to give the first insight into the mode of action of JH, and future studies on neuroendocrine regulations of metabolism will focus frequently on the fat body as a target tissue because of its central role in metabolism and homeostasis.

The fat body has advantages over vertebrate tissues as a model for studying general biochemical processes. The fat body is structurally simple, yet a metabolically complex tissue that shows the varieties of developmental and regulatory processes inherent to higher animals. Because the fat body is simple in organization it is particularly amenable to *in vitro* studies. For example, the presence of an open circulatory system in insects means that hemolymph percolates around and through the fat body lobes, and the adipocytes are arranged for maximum exposure to the circulatory fluid. Hence, suspension of fat body in a culture medium is similar in arrangement to the *in vivo* situation and permits exposure of the cells to the medium with a minimum of artifact. This is unlike vertebrate organs which require perfusion through the complex, closed circulatory system or exposure of the cells to the medium after destruction of the tissue by slicing, homogenization or cell dispersion.

Like vertebrate cells, adipocytes go through stages of growth and differentiation and are sensitive to hormonal regulations. Unlike vertebrates where phases of cell growth and differentiation are gradual and imperceptibly change from one stage to the next, the insect cells reach a stage of development, remain static, then synchronously shift to the next stage in response to endocrine stimuli as the insect molts. These shifts in growth and differentiation are easily observed externally and timed by the act of ecdysis. Hence, events of adipocyte differentiation can be precisely timed for experimentation. This easily observed synchrony is especially useful for studying endocrine regulations of developmental events *in vivo*.

The biochemical processes associated with the development, regulation and function of the fat body will be increasingly significant as they become better elucidated, not only to those interested in insects, but also to developmental and regulatory biologists. The fat body is an ideal model for research on general biochemical processes in higher eucaryotic animals.

INDEX

A

B

C

D

E

F

G

H

N

O

P

U

V

X

Z